THE
AQUAPONIC
FARMER

ADRIAN SOUTHERN & WHELM KING

THE
AQUAPONIC
FARMER

A COMPLETE GUIDE TO BUILDING AND OPERATING A
COMMERCIAL AQUAPONIC SYSTEM

new society
PUBLISHERS

Cover design by Diane McIntosh.

Text Editor: Valley Hennell. Graphic designer: Andrej Klimo. www.andrejklimo.com
Cover photo of plated trout: Our trout served at the Old Firehouse Wine and Cocktail Bar in Duncan, BC (photo credit: Cory Towriss). Important graphic: Adobestock_67829647. All images are author-supplied unless otherwise noted.

Printed in Canada. First printing October, 2017

Inquiries regarding requests to reprint all or part of *The Aquaponic Farmer* should be addressed to New Society Publishers at the address below. To order directly from the publishers, please call toll-free (North America) 1-800-567-6772, or order online at www.newsociety.com

Any other inquiries can be directed by mail to: ´
New Society Publishers
P.O. Box 189, Gabriola Island, BC V0R 1X0, Canada
(250) 247-9737

LIBRARY AND ARCHIVES CANADA CATALOGUING IN PUBLICATION

Southern, Adrian, 1982-, author
The aquaponic farmer : a complete guide to building and operating a commercial aquaponic system / Adrian Southern & Whelm King.

Includes bibliographical references and index.
Issued in print and electronic formats.
ISBN 978-0-86571-858-6 (softcover).--ISBN 978-1-55092-652-1 (PDF).--
ISBN 978-1-77142-247-5 (EPUB)

    1. Aquaponics. 2. Aquaculture. I. King, Whelm, 1977-, author II. Title.

    SB126.5.S68 2017   635'.048   C2017-904850-3
                                                      C2017-904851-1

Funded by the Government of Canada | Financé par le gouvernement du Canada

New Society Publishers' mission is to publish books that contribute in fundamental ways to building an ecologically sustainable and just society, and to do so with the least possible impact on the environment, in a manner that models this vision.

Contents

To all farmers

Acknowledgments

Fɪʀsᴛ ᴀɴᴅ ғᴏʀᴇᴍᴏsᴛ, we thank our primary collaborators on this project: our mothers and editors, Valley Hennell and Kerrie Talbot; Andrej Klimo, graphic designer; Michael Timmons, content reviewer; and Rob West and the whole New Society team.

We thank the pioneers who paved and continue to pave the way: Mark McMurtry and James Rakocy for their foundational work in aquaponics; Michael Timmons and James Ebeling for their voluminous research into recirculating aquaculture and aquaponics; and numerous other farmers who have shared information and ethics with us, directly or indirectly.

Adrian Southern:

I would like to thank my friend Kirsti for planting the very first seed that eventually grew into this book, and my entire family (especially my wife Kim) for their ongoing support during this project. Also, Steve, Janet and Amanda at Taste of B.C. for assisting me in all things aquaculture and answering my endless questions.

Preface

A World Without Weeds

IT ALL STARTED WITH A WEED. It wasn't particularly different than its numerous kin. It was just an ordinary weed, nestled in my rows of neatly planted lettuce, mocking me. It was both benign and the bane of my existence, the cause of the quite literal pain in my ass. I had been SPIN farming for the past two years, using two of my neighbors' backyards to produce a variety of vegetables that I sold at local farmers markets. The work was constant and intense. Converting city yards into small fertile farms was laborious, and managing several of them was a daily struggle. All work considered, I calculated I was earning about $2 per hour. My body was aching, I was just managing to keep my crops reasonably healthy, and there I stood, looking at the weed under a boiling sun — the weed that hadn't been there just a few days ago when I had last spent hours weeding this plot. And it wasn't alone. There was a veritable army of them. As I bent over to dig in yet again, I knew there had to be a better way.

There is.

In 2009 a friend of mine who was enrolled in the Fisheries and Aquaculture program at Vancouver Island University in Nanaimo, where I had lived for some years, invited me to take a tour of the facility. The school had recently set up a small aquaponics system as a demo for the concept. It was a moment of epiphany that would change my life. I was immediately hooked. Raising both plants and fish. Sustainably. All year round. With water use cut by 90% or more. Without the need for arable land.

With. No. Weeds.

After visiting the university, I knew my days as an urban soil farmer were over. For the next three years I voraciously researched aquaponics. I read everything I could find on the subject. I designed and built numerous backyard systems. I experimented and tested. I succeeded and I failed. I became more and more convinced that aquaponics has a vital place in the future of farming.

In 2012, I purchased a property in the rolling hills of the Cowichan Valley on southern Vancouver Island, British Columbia, with the intention of establishing a commercial aquaponic farm. I approached my good friend, Whelm King, an entrepreneur and business manager, to assist me. Together, we created Raincoast Aquaponics (RCA).

Today we grow a wide variety of vegetables and raise rainbow trout in our 36′×80′ greenhouse. Annually, we produce approximately 30,000 heads of vibrant, delicious lettuce (or equivalent other crops) and 750 kg of tender pink trout. We also raise pigs almost entirely on compost and produce fish fertilizer that we bottle and sell to local farmers and gardeners.

A world without weeds is not possible. A farm without weeds is.

Adrian Southern
October 2016

Introduction

The State of the World

As you have just started reading a book on aquaponic farming, we're going to make some basic assumptions. We're going to assume that you understand the urgency of climate change and are familiar with such terms as "peak oil" and "sustainability" and "localization." We assume that you don't need convincing that industrial agriculture is, by its very nature, a system of increasing costs and decreasing returns which turns arable land, one of humanity's greatest resources, into sterile landscapes requiring constant chemical fertilization. The fertilizers themselves are derived from fossil fuels, a dwindling and polluting resource.

Industrial agriculture has disrupted the natural methods of farming that have sustained humans for millennia. It produces low-quality food heavily depleted of the essential elements necessary for human health. Fertile land becomes barren, human health deteriorates, and the whole system requires vast infrastructures to grow, store, move, store again, move again, store yet again and so on, before it is finally sold to us in all its nutrition-lacking glory. The whole system is fragile and rigid, every link in the chain essential and requiring large inputs. If even one link breaks, all efforts are spoiled and all food wasted. In permaculture terms, the system lacks any semblance of redundancy.

Industrial agriculture is inherently unsustainable, and the system is breaking down. Global food supply is increasingly unstable with food prices sharply increasing in many parts of the world. Here in North America this reality has been mostly hidden due to government subsidies.

Once in a lifetime droughts are now common. Pollinator colonies are collapsing. Super weeds, resistant even to the poisons that created them, are rampant. The industrial promise of low food prices is being revealed as the sham it always was.

We continue to rely on industrial agriculture at our own peril. Change is required.

In summarizing our food system in this manner, we assume we're preaching to the choir. We assume that you want to be part of the solution — the movement to reclaim our food systems — for the sake of both healthy ecosystems and our own health, and to allow future generations the opportunity to survive if not thrive.

The growing movement to counteract the ills of industrial agriculture and globalization is robust and filled with vitality and energy. It is a movement of the people for both the people and the land. It is a movement designed to endure. The central tenet is localization.

Produce locally. Buy locally. Use locally. Support locally. Be local.

Relocalization of food production can take two primary forms: moving backward or moving forward.

Moving backward means using the time-tested methods that have sustained humans since agriculture was invented. It is the revitalization of traditional, small, labor-intensive organic farms. It is nurturing the land and managing natural ecosystems, creating soil teeming with microorganisms and farming in harmony with and within the limits of local environments. It is an ancient system whose flag might best be represented as a shovel and compost pile.

Moving forward is using technological advancements and scientific knowledge to produce food outside of natural ecosystems, virtually anywhere it is needed. It is using resources and ingenuity to create our own ecosystems to produce food with almost no environmental impact, in almost any climate. It is building the capacity to produce food locally in all seasons with highly efficient labor and water use. We believe aquaponics is moving forward.

We are advocates for both moving backward and forward. These methods are not in competition: both have advantages and disadvantages and are vital to food sustainability. We have the utmost respect for traditional farmers. We have chosen to be pioneers. We are aquaponic farmers. Join us!

What Is Aquaponics?

A Primer on Aquaponics

THE WORD "AQUAPONICS" was coined in the 1970s as a combination of the words "aquaculture" and "hydroponics." Aquaculture is the cultivation of aquatic animals and plants in natural or controlled environments. Hydroponics is the growing of plants without soil, using water to carry the nutrients. The term "aquaponics" was created to designate the raising of fish and plants in one interconnected soilless system.

Aquaponics can solve the major problems of both freshwater aquaculture and hydroponics.

The major problem in land-based aquaculture is that fish waste in the water creates continuously elevating levels of ammonia. If left unchecked, this toxic element will rapidly kill the fish. The aquaculture industry typically uses one or both of two options to resolve this problem: a constant supply of fresh water to replace the toxic water and/or expensive filtration systems. Neither is ideal. The former not only uses voluminous quantities of our precious fresh water but also creates equally large quantities of high-ammonia water that is toxic to any natural ecosystem. The latter is simply very expensive. The high cost is especially pertinent to smaller commercial operations as most filtration units only make financial sense at large economies of scale.

Fish farms in natural bodies of water, often called "open net pens," are rife with problems, notably their potential for negatively impacting wild fish stocks. We do not support such farms, and they are not considered in this book.

The major problem in hydroponics is the ongoing need for large inputs of fertilizers. A soilless production system means all the minerals — all the food — required by the plants must be continually added. Fertilizers are expensive, and the vast majority are fossil-fuel derived, often referred to as "chemical" fertilizers. Available organic fertilizers are not commonly used because they are less water soluble, thus more likely to cause problems and can be several times more expensive than their chemical counterparts. Hydroponic farms are often also a major water consumer as many use a drain-to-waste system.

The aquaponic cycle.

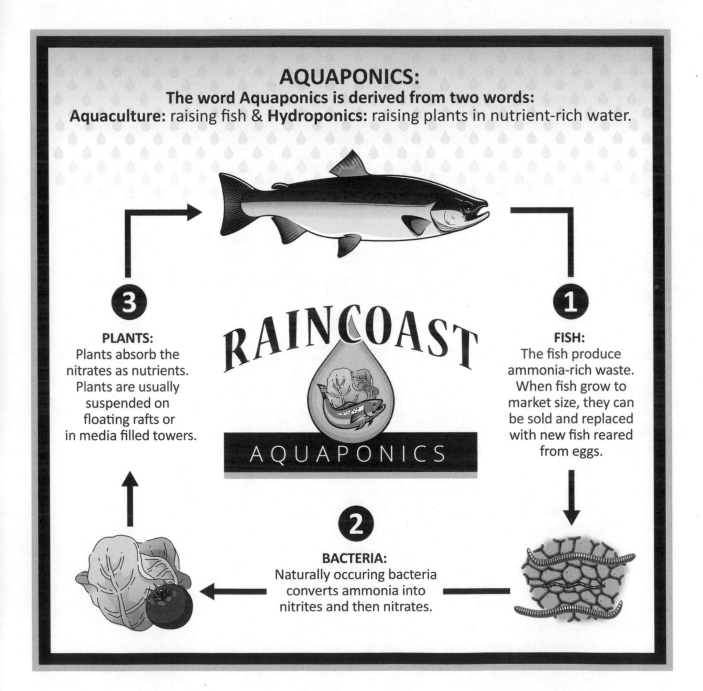

AQUAPONICS:
The word Aquaponics is derived from two words:
Aquaculture: raising fish & **Hydroponics:** raising plants in nutrient-rich water.

RAINCOAST
AQUAPONICS

③
PLANTS:
Plants absorb the nitrates as nutrients. Plants are usually suspended on floating rafts or in media filled towers.

①
FISH:
The fish produce ammonia-rich waste. When fish grow to market size, they can be sold and replaced with new fish reared from eggs.

②
BACTERIA:
Naturally occuring bacteria converts ammonia into nitrites and then nitrates.

Even hydroponic farms that recirculate water must drain and replace their water regularly as they do not host a living ecosystem that balances itself.

By combining fish and plants into one system, aquaponics can solve the primary problems of both aquaculture and hydroponics. Fish waste provides a near-perfect plant food and is some of the most prized fertilizer in the world. The plants, using the minerals created from the waste, do most of the work of cleaning the water for the fish.

The fish feed the plants. The plants clean the water. The symbiosis is as logical as it is effective.

The third living component in aquaponics is bacteria. The whole system hosts specific types of bacteria that serve two roles. One family detoxifies ammonia in the effluent by converting it into nitrates. Another family mineralizes organic material (primarily fish feces and uneaten feed) by breaking it down into its elemental constituents, which are usable by plants. Without this vital conversion in a closed system, both fish and plants would rapidly die. Establishing the bacterial cultures and monitoring their health is one of most important tasks of an aquaponic farmer. We cover this topic in depth in Chapter 6.

A Very Brief History of Aquaponics

Although modern aquaponics is only a few decades old, the concept of combining fish farming and plant production for mutual benefit is thousands of years old.

Since ancient times, fish have been raised in flooded rice paddies in China. The fish and rice are harvested at the same time annually, and the technique is still used today. Ducks, sometimes in cages, were kept on the edges of fish ponds so their excrement could be used to feed the fish.

The Aztecs had advanced techniques of aquaponic farming called *chinampas* that involved creating islands and canals to raise both fish and plants in a system of sediments that never required manual watering, achieving up to seven harvests per year for certain plants.

In 1969, John and Nancy Todd and William McLarney founded the New Alchemy Institute in Cape Cod, Massachusetts, and created a small, self-sufficient farm module within a dwelling (the "Ark") to provide for the year-round needs of a family of four using holistic methods to provide fish, vegetables and shelter. In the mid 1980s, a graduate student at North Carolina University, Mark McMurtry, and Professor Doug Sanders created the first known closed loop aquaponic system. They used the effluent from fish to water and feed tomatoes and cucumbers in sand grow beds via a trickle system. The sand also functioned as the biofilter of the system. The water percolated through the sand and recirculated back to the fish tanks. McMurtry and Sanders' early research underpins much of the modern science of aquaponics.

The biggest leap came from Dr. James Rakocy at the University of the Virgin Islands. From around 1980 through 2010, he was Research Professor of Aquaculture and Director of the Agricultural Experiment Station, where he directed voluminous research on tilapia in warm-water aquaponic systems. His research on the conservation and reuse of water and nutrient recycling remains the greatest body of modern work on aquaponics. Though it took many years to develop, by around 1999 Dr. Rakocy's system had proven itself to be reliable, robust and productive. His developments are used today from home to commercial-scale aquaponics.

Our work has been primarily developing systems and protocols that have allowed us to modify the work of such visionaries as McMurtry and Rakocy to cold-water production, better suited to colder environments.

Aquaponic Ecomimicry

Ecomimicry is the design and production of structures and systems that are modelled on biological entities and processes. Aquaponic systems are manufactured environments that attempt to replicate a complex natural system. Every component and process in an aquaponic system has a natural counterpart.

Imagine a freshwater ecosystem. At a high elevation is a lake in which fish constantly produce waste in the form of ammonia and feces. A river flows from the lake carrying these wastes. Along the bottom of the river are layers of gravel and sand which are home to various bacteria and invertebrate detritivores (worms, insects, crayfish, etc.)

As waste-laden water flows down the river, feces sink to the bottom and are trapped in the gravel where it is eaten and broken down by detritivores and bacteria, converting it into elemental constituents and minerals. Ammonia (a toxic form of nitrogen) in the water is nitrified into nitrates. Without bacteria and detritivores, the waste would eventually build to toxic levels.

The river continues downstream to lower elevations and eventually meets a wide, flat wetland area. Here it slows and spreads out, depositing mineral-rich sediments where vegetation abounds.

After being filtered of its nutrients and sediments in the wetland, the water ends its downhill journey in the ocean. But this is not its end. Evaporation and evapotranspiration from plants combine to form clouds, and their moisture falls as rain, which collects in large bodies of water such as lakes, and the cycle repeats.

All these natural processes are found in an aquaponic system: the fish tanks are the counterpart to the lake, the filtration systems are the gravel in the river, and the hydroponic subsystem is the wetland. The main water pump serves as clouds by returning the water to the high point in the system: the tanks.

As we are mimicking a natural ecosystem, many challenges found in an aquaponic system are also found in nature. Nature had billions of years to evolve solutions which may be replicated in aquaponic farms by imitating nature.

Aquaponics, Permaculture and Sustainability

We believe aquaponics is a system of permaculture. All three tenets and twelve principles of permaculture design are realized within an aquaponic system, from conception and design to operation.

One of the core tenets of permaculture is the "return of surplus" which is maximizing the efficient use of resources and eliminating waste. Often, waste can be eliminated simply by recognizing it as a resource and using rather than discarding it. An aquaponic system has this tenet at its core, as observed in the relationship between fish, bacteria and plants.

Aquaponics has inputs and outputs. When permaculture design principles are applied, the inputs are minimized and used efficiently and the outputs are recycled back into the system as inputs. At Raincoast Aquaponics, we extract five different uses from every kilogram of fish feed and three uses from every liter of water.

The fish feed is used to raise fish (1), which in turn feed plants (2) via the bacteria. The resulting fish waste is captured and converted to a fertilizer product (3), and the crop residue (compost) is fed to pigs and converted into bacon (4). Pig waste is composted and used to build soil for growing field crops (5).

Water is first used to purge fish prior to harvest (1), and then used to top up the main system (2). The effluent flushed from the system is used to water field crops (3) after most of the fish waste has been extracted.

Aquaponic Plant Systems

There are several commonly used aquaponics systems whose names refer to the method of plant production. Systems of raising fish are all very similar, thus not considered in naming aquaponic systems. In all systems, two basic functions are found: water is cycled between the fish and the plants, and bacteria convert fish waste to beneficial minerals.

The four most commonly used aquaponic plant production systems are: Deep Water Culture, Drip Towers, Nutrient Film Technique and Media Bed.

Deep Water Culture (DWC): water flows down long troughs of water, typically about 12″ deep, like a slow-moving stream. Rafts, typically made

DWC troughs with floating polystyrene rafts.

from styrofoam, float on the water with a pattern of holes cut into them. Small open-bottom pots, called net pots or slit pots, fit into the holes. Plants are supported in the pots by a variety of different mediums. The roots of the plants are suspended and grow in the moving water.

Drip Towers are tubes, typically made from PVC, with either holes or a slit running the length of the tube on one side, suspended vertically in rows. The towers contain a growing media into which plant roots grow. Water is continuously fed into the top of each tube and collected at the bottom to cycle through the system again.

Nutrient Film Technique (NFT) also uses tubes, typically PVC, with holes on one side. Whereas drip towers are suspended vertically, NFT tubes are mounted horizontally on a slight angle with the holes facing upwards. Plants are grown in small net pots inserted into the holes in the tubes. Water, continuously fed into the high side of the tubes, flows down in a thin film contacting the roots and is collected at the low side to cycle through the system again.

Media Bed is a type of flood and drain production with numerous possible configurations. In all configurations, watertight growing areas are flooded at regular intervals by pumps then drain back to cycle through the system. The growing areas are filled with a pebble-like medium, often expanded clay aggregate but sometimes simply gravel. Plant roots grow throughout the medium.

While there are pros and cons to each system, in our opinion the only two systems worth considering for a commercial operation are DWC and drip towers. Media bed systems are not practical due to the maintenance required to remove trapped solids and the higher risk of rapid crop failure if a mechanical problem occurs. NFT systems are widely used in hydroponic production but are inferior to drip towers and DWC for both bacteria colonization and space usage. We strongly suggest DWC or drip towers for commercial production. This book is based around a DWC system.

Healthy roots under a DWC raft.

ZipGrow™ Towers.
CREDIT: BRIGHT AGROTECH

ZipGrow™ Tower array.
CREDIT: BRIGHT AGROTECH

Deep Water Culture Systems

The primary advantages of DWC are:

- Less expensive to construct
- Even light distribution
- Increased thermal mass due to the large volume of water in the system
- Ability to selectively move individual plants for thinning and spacing
- Greater options for pest control

Less expensive:

Cost is by far the biggest advantage of DWC. For a 120′×36′ greenhouse, you can expect to spend at least US$50,000 more to install a drip tower system compared to DWC.

Light distribution:

All plants in a DWC system have relatively equal access to light. As they are all on one horizontal plane, they are potentially only partially shaded by their immediate neighbors. In contrast, the vertical design of a tower system means an increased potential for shading, particularly for the lower plants and all the more so when using supplemental light that will not penetrate as effectively as sunlight.

Thermal mass:

DWC systems contain about three times more water than a drip tower system due to the volume of the troughs. A DWC system in a 120′ greenhouse will have approximately 66,000 liters. A tower system will have approximately ⅓ this volume (18,000 L). The additional water serves as thermal mass which buffers temperature during cold and hot periods, and does so where it is needed most, immediately around the plants.

Thinning and spacing:

Rafts have both advantages and disadvantages. A primary advantage is that individual plants can easily be removed from the system or relocated which allows you to start plants closely spaced and spread them out as they grow.

Pest control:

It is also easy to see and remove problem plants such as those with pests, mold or mildew. Additionally, plants in a DWC can be directly washed with system water without losing the water. This is a highly effective method for maintaining plant health and is not possible in a tower system without losing the water used.

The disadvantage to rafts is that they are near ground level (approximately 1′ above ground), thus the work involves regularly bending over. Fully loaded rafts can weigh 30–40 lbs. and can be awkward to move, though carrying rafts with mature plants over long distances is not needed or recommended in our system as most harvesting takes place at the troughs.

Drip Tower Systems

All mentions of towers or the performance and layout of towers in this book refer to ZipGrow™ Towers by Bright Agrotech. These are the only towers we specifically recommend.

The primary advantages of drip towers are:

- Potentially increased numbers of plants per square feet of growing space
- Increased nitrifying capacity, as the tower media has a large Bacterial Surface Area
- Increased oxygen availability to plants due to the high porosity of the media
- Potentially easier workflow

Increased plant sites:

Income from an aquaponic system is mostly made from plants, not fish, so an increase in plant sites directly corresponds to an increase in potential income.

Increased Bacterial Surface Area (BSA):

BSA is the area within a system that nitrifying bacteria can colonize. Large BSA increases are only found in matrix-based tower systems such as ZipGrow™ Towers. The superior BSA due to the media used in ZipGrow™ Towers is a substantial advantage. A 5′ drip tower can provide approximately 150 square feet of BSA in the tower media. DWC without additional biofiltration provides approximately 6 square feet of BSA in the same space.

Increased oxygen availability:

In DWC systems, plant roots are submerged under water which limits the types of plants suited to the system. Supplemental oxygen must be added by an aerator, and its availability to the plants is limited by the oxygen carrying capacity of water which is approximately 10 ppm at 15°C.

With drip towers, roots grow into a porous media through which water is slowly trickled. Rather than being submerged in water, roots are directly exposed to the air and thus the oxygen in the air is available to the plants. Tower systems are self-aerating due to the high air/water exchange as water trickles through the media.

Easier workflow:

The workflow in both systems is simple once practiced. The primary difference is that in DWC all the plants are located at the same height (about 1′ off the ground). In a tower system using 5′ towers, plants range in height from about 1′ to 6′ and the towers can be easily carried around.

DWC or Drip Towers — Our Recommendation

It is important to understand that if properly designed, constructed and operated, both DWC and tower systems will produce beautiful plants and high-quality fish. We use DWC so admit our bias. We acknowledge that towers have some distinct advantages and recommend them as an excellent production system, but overall we believe DWC is superior.

Plant Sites and Light Availability

While drip towers have an advantage in plant sites, the difference between the two systems is not as great as it might seem. A 120′×36′ greenhouse with 86′ of plant production area has the potential for 8,000 plant sites using 5′ drip towers and 6,192 harvestable plant sites using DWC (see Chapter 2).

The increased plant capacity for towers (8,000 vs 6,192) would seem like a game changer, but we do not agree. The 8,000 number is based on the recommended spacing for ZipGrow™ Towers of 20″ between plant centers side to side and 16″ centers back to front, with 36″-walkways between arrays.

Our concern is that, in a temperate latitude or colder, light will not sufficiently reach the lowest plants. Our concern is doubled during the 4–6 months a year when lighting supplementation is required. In contrast, in a DWC system the plants are all at one horizontal level, thus access to light is virtually identical throughout the greenhouse and light supplementation is evenly distributed. So while towers have a theoretical advantage in plant sites, we feel the limitations on light distribution greatly reduces if not eliminates this supposed advantage.

We also note that to house 8,000 plant sites requires 4 arrays with 5 rows of towers per array in a 36′-wide greenhouse. This means the walkways are only 28″ which will further reduce light penetration and creates a tight working area. Note that this spacing leaves walkways that are narrower than the 36″ recommended by ZipGrow™ for light penetration and working space. If you reduce to 3 arrays, you can widen the walkways to 48″ which increases light penetration but also reduces the total plant sites to 6,120.

Another option is to use 3 arrays with 6 rows per array instead of 5. This allows for 7,152 plant sites and 35″ walkways with rows 18″ on center. The downside of this option is that it may be problematic for the plant production schedule, which involves rotating towers through an array as plants mature. A 40′-wide greenhouse will help solve this problem by increasing walkways to about 38″ with 4 tower arrays.

Ultimately, we feel that any gain in plant sites in a tower system is countered by a lack of equal light, or that providing equal light means virtually the same number of plants sites as in DWC.

Bacterial Surface Area

The greatly increased BSA of a tower system that uses a matrix media is a real advantage. However, it is important to understand that the bacterial colony capacity of DWC is more than sufficient to raise healthy fish and excellent plants. Additionally, BSA can be increased in any system with the use of a biofilter module, as we have done in our design.

From a functional standpoint, the extra bacterial capacity of drip towers means that you will have more wiggle room in terms of the total fish load in the system and you can be less precise with the quantity of fish feed (excess fish feed also breaks down into ammonia). In other words, the primary advantage of the increased BSA is not increased plant production, which is where your profits come from, but in increased fish capacity and in needing to be less precise with how you run your system. Note that the increased BSA only applies to matrix-media based systems such as ZipGrow™.

Available Oxygen

While greatly increased oxygen is a real advantage for drip towers, the oxygen levels in DWC are more than sufficient to raise most types of plants rapidly. So while the advantage is clearly to drip towers in oxygen capacity, both systems will excel if designed and run properly.

It should also be noted that the same characteristics that allow towers to be self-aerating also cause the water to gain or lose heat quickly due to its exposure to the air. Hence tower systems have much less thermal stability than DWC systems.

Filtration

The media used in towers act as thousands of filters. The filtration capacity in a tower system is unparalleled. That said, as with most other tower advantages, the filtration in our DWC design is more than sufficient to produce at levels of high efficiency in volume over the long run. The advantage is to towers, but it is less important than it seems.

Our Conclusion

All things considered, we feel the advantages of a tower system are considerably less than they first appear when compared with our DWC system. When you factor in the advantages of DWC, notably the lower cost of construction and the consistency of light availability, we feel DWC is the superior system.

To be clear, we approve of and recommend tower systems such as ZipGrow™ for aquaponic production. They have been proven to excel. We do, however, feel the benefits compared with a well-designed DWC system are less than tower proponents claim and that the benefits do not justify the considerable extra cost, unless cost is not a concern for you. Though this book is based on a DWC design, we will at appropriate times throughout convey information specific to tower systems.

Backyard vs Commercial Systems

Backyard systems typically cannot be scaled up to commercial systems of the size discussed in this book. We have constructed numerous backyard systems over the years. They are a fun project that can be very productive, and we highly recommend them to everyone with the available space and basic building skills. The differences between a backyard system that might contain as little as 100 liters of water versus a 120′ DWC system that contains 600 times that volume are considerable.

In a small system that might occupy less than 10 square feet, most components can be made quite simply or easily sourced at the local hardware store. Problems in the system such as leaks are easy to see, easy to fix, and the repercussions of a failure are small.

In a commercial system that will likely cost hundreds of thousands of dollars, where leaks might not be visible as much of the water pipe is buried, and where problems can lead to losses of whole fish cohorts or plant crops worth many thousands of dollars, the design of the system has to be commensurate to the risk. Additionally, the scale of a commercial system requires entirely new components, such as a large particle filter (we use a Radial Flow Separator — see Chapter 2), UV sterilizers and a waste collection system.

While all aquaponic systems share the same basic parameters such as the cycling of water and the fish-plant ratio (see "The Golden Ratio" in Chapter 3), the design of commercial systems is different from backyard systems.

The RCA System

2

IN THIS CHAPTER we dive into the practical aspects of aquaponic systems in general and our system specifically. This chapter introduces all the major elements of an aquaponic system, including the property; the greenhouse and the major components, water management; and the live entities in the system: fish, plants and bacteria. This is the most important chapter to read to gain an understanding of aquaponics and of our system specifically.

The Purpose of This Book

While we have made assumptions about your understanding of the state of the world, we make no such assumptions about your knowledge of aquaponics. We intend this book to be sufficient for someone who has never heard of aquaponics to understand all its primary elements, from the science behind the system to the nuts and bolts of setting up and operating a commercial aquaponic farm.

We farm in southwestern British Columbia, on Vancouver Island close to the Pacific Ocean. Our farm is in an 8A hardiness zone. This book is recommended for aquaponic farms in temperate regions or cooler.

Much of our design work has been the adaptation of the University of the Virgin Islands tropical system to work efficiently and sustainably in colder regions. The primary differences are operating at cooler water temperatures and using locally adapted fish species as well as appropriate plant varieties.

There are several plant production methods that are possible in an aquaponic farm. This book is based around a Deep Water Culture (DWC) system, with some auxiliary guidance for tower-based systems.

This book is also designed around building the entire aquaponic system within one greenhouse. Alternatively, you can build a separate building to house the aquaculture subsystem. If you opt for a separate aquaculture building, all the core design principles and work processes apply, but the layout and in-ground plumbing will change.

In our experience, the minimum size of greenhouse that should be considered for a commercial operation is 120′ long by 36′ wide. This size has the capacity to generate an excellent return on investment yet can be easily operated by 1–3 people at less than 40 hours combined farming work per week (not including sales, marketing and administration). We have written this book around this size of greenhouse though the formulas and design principles can be scaled up to systems many times larger.

Many of the design principles are not relevant or recommended for backyard or home-based systems. For those interested in small non-commercial systems, we strongly recommend Sylvia Bernstein's excellent book *Aquaponic Gardening* (New Society Publishers, 2011).

While numerous types of plants thrive in an aquaponic system, the plant we use throughout the book is head lettuce. Using head lettuce as a unit of production has numerous advantages for both design and business planning:

• Its production time is definable
• It takes a specific size of area to grow at different stages of production
• Sales are measured in units, not weight
• Price estimates are easy to establish based on different sales outlets

In other words, head lettuce allows us to easily play with all the key variables to determine the growing process and sales potential. When we refer to "plant sites" throughout the book, they are sized for head lettuce but are easily adaptable to most plants.

This book is also based around the raising of rainbow trout. Trout have an excellent feed conversion ratio, are typically easy to source as fingerlings and are highly marketable. While other cold-water species can be raised in the system if desired, you should do additional research in advance to confirm suitability. Notably, some other species that are considered cold-water, such as channel catfish or sturgeon, may prove to be advantageous in cold-water aquaponic systems though they have not yet been tested by us.

In summary, if you build an aquaponic farm based on the design in this book, you will be building a cold-water aquaponic facility in one 120′ greenhouse that houses all production systems (except the Germination Chamber), that is capable of producing up to 80,000 heads of lettuce per year in a Deep Water Culture (DWC) system, raising 750 rainbow trout per year to approximately 1 kg each, and that can be operated by 1–3 people working less than 40 hours combined per week.

Avoiding Our Mistakes

By following the system laid out in this book, you will be avoiding numerous problems that come with designing your own system. You will be bypassing the many mistakes and system idiosyncrasies that we encountered and overcame in creating the system presented here. For example, over the first 18 months, we lost four cohorts of fish for four different reasons, each of which led to improvements and innovations. It was a frustrating beginning that we hope to help you to avoid.

One loss was due to excess ammonia in the initial cycling, due to our biofilter not being fully active because of a lack of calcium in the water. One loss was due to fish pathogens and convinced us of the importance of UV sterilization and fish treatment methods such as salt baths. One loss was due to a power outage and led to our backup oxygen system. And one loss was due to a wayward frog entering our sump and seizing up the single pump we initially used. That frog led to our two-pump design and the use of a monitoring system.

The systems in this book are the result of countless lessons learned the hard way, minor and major tweaks, and vast amounts of research and innovation. They also incorporate commercial aquaculture design and components in contrast with the "do-it-yourself" mentality that is common in aquaponics. The resulting design uses widely available, mostly low-tech equipment in a system that is proven, predictable and high-performance.

A Note on Reading Before Building

Before seriously considering constructing an aquaponic system, we suggest you read through this entire book. While it is laid out in a logical format, from early site considerations through marketing of products, you should be familiar with the totality of the information before proceeding to break ground and invest in a facility.

Note as well that our farm is a prototype. Some of the images presented are slightly different than the design in this book but are shown for demonstration purposes. Follow our instructions and diagrams closely.

A Note on Metric vs Imperial

All construction measurements throughout the book, including pipes for plumbing, are imperial (inches and feet). Construction in North America is still generally measured imperially.

All aquaculture measurements throughout the book, notably water volume and flow, are metric (centimeters and meters). Within the aquaculture industry, metric is standard.

We have tried to simplify and clarify this contradiction as much as possible, though we recognize it cannot fully be remedied as the construction industry still operates in imperial and the aquaculture industry in metric.

A Note on Currency

All sale prices and cost estimates throughout the book are given in US dollars unless otherwise specified. Some large figure amounts are marked US$ for emphasis. Canadian currency is marked as C$. Figures given are to the best of our knowledge as of the date of publication.

A Note on North

References to compass directions assume your greenhouse is situated with north as indicated in Diagrams DWC OVERVIEW and TOWER OVERVIEW on pages 30 and 31, which allows us to easily reference locations within the greenhouse. For example, this means your sump will be on the south side of the greenhouse. Your Seedling Table will be in the northwest corner. If your greenhouse is situated differently, adjust according to your site.

Property Considerations

There are several qualities to consider in situating your greenhouse. The main considerations are:

- Zoning
- Sun exposure
- Characteristics of the land
- Access to power and water
- Prevailing winds
- Waste disposal
- Long-term land rights
- Living onsite

Zoning

Not all locations will be legally zoned to allow aquaculture and/or agriculture. Before any other consideration, confirm with your municipality and governing fisheries body that the property is acceptable for aquaponics.

Sun Exposure

The rule here is: the more sun the better. Sunlight is the primary source of energy for your plants, and there is no such thing as too much. It is always possible to shield out the sun (see Shade Cloth in Chapter 5), but the location of your greenhouse usually sets a firm upper limit on sun hours and intensity.

Pay special attention to how early and late in the day the sun hits the potential site. Do you lose the sun early in the evening because it goes behind a neighboring mountain, or does the sun hit the site later in the morning because of a grove of trees to the east? If possible, locate your greenhouse where the earliest and latest sun of the day will reach your plants. We recommend doing a sun chart to gauge this. Chart for the winter solstice as this will tell you the least sun during the year. The duration of sun hours over the winter will determine the amount of supplementary light required for year-round production.

As a guide, lettuce requires a minimum of 14 hours of strong light per day to produce finished heads in five weeks in a trough or tower (sprouting and seedling stages are additional). Plants will continue to grow with less sun hours and/or less direct sun exposure, but the production time can greatly increase, which means not only less plants to sell but also increased potential for disease and pest damage.

In almost every situation, the greenhouse should be oriented east-west. In the case of a drip tower system, this is mandatory for effective light penetration. If the topography of the land or extreme prevailing winds prevent you from orienting your greenhouse east-west, a DWC system can be oriented in any direction with only minimal loss of light penetration.

Characteristics of the Land

The primary characteristics to consider are grade and constituency. The grade of the land is simply the slope, or angle, of the land. Your greenhouse will need to be totally flat so that water moves in the direction you want it to. A site that is nearly flat will save a lot of time and money in grading the land to level.

The constituency of the land is the earth itself, typically sand, clay or loam (a mixture of sand and clay). While a sandy site will be the easiest to work with, the most important factor is compaction. The earth will be supporting the weight of many tons of water, particularly under the tanks and troughs, so land that is not fully compacted will tend to settle under the weight and cause problems over time.

Access to Power and Water

Aquaponics relies on electrical power. It is not optional. In this book we are assuming you will be using power from the grid. Accordingly, an early assessment by an electrician of the costs of bringing dedicated power to the greenhouse is vital. If you are powering your greenhouse via off-grid means located onsite, all the same needs apply except for the grid tie-in. Even if using off-grid power, we strongly suggest being connected to the grid as power outages can be fatal to an aquaponic farm.

Access to suitable water is also crucial. We suggest testing your water source while assessing a proposed location. Water for the system should be clean, clear, free of pathogens and bacteria, and have a pH between 6.5 and 7.5. Always have your water tested for biological and mineral content (see Chapter 6).

Typical sources for water are the same as those used in households: a municipal supply, a deep well or harvested rain. Well water that has not undergone chemical treatment is ideal. Municipal water can be used as long as it is free of chlorine and chloramines. Most municipalities disinfect their drinking water with chlorine, which can be easily removed by vigorously aerating it for a few hours or letting it stand in an open container for at least 24 hours, though this extra purification process is inconvenient.

It is important to note that chloramines cannot be removed from the water using the aeration or 24-hour method. Chloramines are removed by UV light, which is part of our system to clean and sterilize the water. Our system design will actively remove chlorine and chloramines by design, but we recommend that if you are using municipal water, you install an additional small UV pre-filter on the main water line to the greenhouse.

We strongly suggest not using surface water from a natural body such as a river, lake or pond. Doing so may bring in any number of microbial problems and may not be allowed by your local government without a water license. Source water is so important that we suggest seeking a different site if your only option is to use surface water. If you must use surface water, it must be sterilized and filtered to potable quality prior to entry into the system.

Prevailing Winds

Think of your greenhouse as a very large kite. Given the right wind power and direction, huge forces are trying to make it take flight. This should not happen with the proper hardware and installation, but it is a good idea to understand the physics and plan accordingly. If you purchase your greenhouse new from a manufacturer, make sure it is engineered to withstand the wind and snow loads typical of your area. On a standard poly covered greenhouse, the prevailing winds should hit the side wall (one of the long, rounded sides) rather than the endwalls (the vertical walls at the ends).

The potential conflict with this is that if you opt to use drip towers, the greenhouse must be oriented east-west to allow effective light penetration. If you opt to use troughs, the greenhouse can be situated in any orientation without substantial compromise to light capacity.

Waste Disposal

In the normal operation of your aquaponic system, you will be manually cleaning filter pads which have trapped large quantities of fish feces and debris, and flushing out the Radial Flow Separator (RFS) and fish tank drains. This is the only source of water loss from the system other than evaporation and transpiration. The wastewater (effluent) needs to be removed, and in many cases (particularly in Canada), proper effluent disposal will be a condition of your aquaculture license. How you dispose of the effluent will depend entirely on your unique situation.

The simplest method is to distribute the effluent to a leach-field and allow it to percolate into the topsoil. "Ground infiltration" can be done by simply spraying the effluent onto a hayfield or by constructing a dedicated septic-style disposal system.

In our opinion, this is like throwing cash on the ground and waiting for it to soak in because your fish wastewater is actually a form of liquid gold. Aquaponic wastewater is full of all manner of elements, minerals and organic solids that can be extracted and fermented (biodigested) into an amazing fertilizer, which can either be sold for extra profit or recycled back into the aquaponic system to supplement nutrients. The mineral-rich water left over can be used to irrigate a nearby orchard or field crop.

Whatever you choose to do with the effluent, you need to consider how and where you will capture it. The design presented in this book has effluent drains connected to the fish tank standpipes, the RFS and the washdown sink (for cleaning off the filter screens). These drains carry the effluent to the Waste Tanks outside the greenhouse.

On our farm we were able to take advantage of the natural slope of the property and locate Waste Tanks (IBC totes) about 20 feet away from the west endwall, just below the grade of the greenhouse. If your property is perfectly flat or you are unable to take advantage of the topography, you will need to excavate a pit so that the Waste Tanks are lower than the Main Waste Pipe (see Chapter 4).

The Waste Tanks can get very smelly, so locate them an appropriate distance from the greenhouse and any nearby living areas. You will also need to take steps to secure the sump or tank with a lid to prevent any accidents such as children or pets falling in. Whether you use a tank or a waste sump, you will want a vessel that is at least 1,000 liters, which will give you 4–5 days of collection before it needs to be emptied. See Effluent later this chapter.

Long-term Land Rights

Most farmers rightly view farming as a long-term project. Developing the land, often requiring restoration from non-use or poor use, building infrastructure and learning the microclimate and what grows best can take many years and considerable investment of time and labor. Securing long-term land rights, whether through ownership or lease, is highly desirable. With an aquaponic farm, it is mandatory. The infrastructure you build is the farm and should be considered permanent. Moving a commercial aquaponic farm is possible with large expense and effort, but we suggest you only consider becoming an aquaponic farmer if you can secure land rights for a minimum of 10 years.

Living Onsite

It is very advantageous to live on your farm. You can easily check in on things in the greenhouse multiple times per day, you can likely respond to emergencies much more rapidly and it is much more pleasant to have a stroll on your farm rather than a drive as your commute. If possible, plan on living where you farm and source your property to accommodate this.

The Greenhouse

Size

Just like buying a home, buying a greenhouse has a huge array of options and prices. The most important thing to consider is size. The size of your greenhouse is the starting point of the entire design process. This book is based around a 120′ by 36′ greenhouse.

We have designed our system around a 120′ greenhouse for several reasons:

1. The greenhouse houses all necessary components under one structure with no wasted space.
2. The system is within the optimum range of the Golden Ratio (see Chapter 3). If you increase or decrease the trough length by more than about 20′, you will be outside the Golden Ratio and will require differently sized fish tanks and components throughout the system.
3. This size of operation is ideal for a family farm: it can be farmed by 1–3 people working a combined 40 hours per week and produce an excellent income. Note that this estimate of hours will depend greatly on the efficiency of the farmer(s) and is an estimate of farming time only. Administration, marketing and sales are additional.

The work area has been designed to fit all the necessary components as efficiently as possible while maintaining a comfortable working space for all tasks. If you prefer to have more room, you can build a longer greenhouse and expand the work and seedling areas.

If you have the option to acquire a 40′-wide greenhouse without great additional expense, we recommend doing so. The wider walkways between troughs (or towers) and slightly more open space in the work area are worth it.

It is certainly possible to use a greenhouse of different dimensions, such as square, or greenhouses longer than 120′, but you will need to rearrange the layout of both hydroponic and aquaculture subsystems to make sure they fit within the size. Use the formulas in Chapter 3 to ensure that your design is well balanced.

At Raincoast, we operate out of an 80′×36′ greenhouse due to site limitations. Our greenhouse is very productive, but we would certainly expand to 120′ if we had room to do so.

IMPORTANT

Remember that whenever we talk about a greenhouse in this book, including estimated prices, we are referring to a 120′ long structure that is 36′ wide, the minimum size we recommend.

New vs Used

You can be successful with a new or used greenhouse. In the end the decision will likely come down to how much you are willing to pay versus how much work you are willing to do to rebuild and perhaps fix a used structure. Availability and condition are also key factors if buying used.

The biggest advantage of a used greenhouse is price. A new double-layer poly covered greenhouse, which is the most common type, will likely cost around US$25,0000 for the structure itself (not including installation, heating or ventilation). A new twin-wall polycarbonate-covered greenhouse will cost around three times that much. A used greenhouse can often be found for US$10,000 or less, though you may have to deconstruct it from its previous site, which is no small job.

Potential downsides of a used greenhouse are:

- Uncertainty as to the quality of the structure itself.
- Uncertainty as to the snow and wind rating of the structure.
- Uncertainty as to whether all components of the structure are with it. For example, wind bracing may be missing and you won't know. Missing components can greatly reduce structural integrity.

- If the structure was mounted in concrete, which we recommend (see Chapter 4), it may be problematic to remount at your location (note whether the arches are bolted onto ground-stakes or set directly into the concrete).
- Limited capacity to choose features such as roll-up sides.
- Poly covering will most likely need to be replaced immediately.

A new greenhouse will give you a structure that is designed for your needs, engineered for your location and should be warrantied. It will undoubtedly cost more.

Types of Covering

There are three main options for greenhouse covering: polyethylene ("poly"), polycarbonate and glass. Of the three, poly and polycarbonate are the two we strongly suggest you consider.

Glass is by far the most expensive option. Even though glass boasts the highest insulation value, it is extremely expensive to construct and repair, and for this reason alone, we do not recommend it.

Single-layer poly consists of a single layer of thin, flexible polyethylene plastic. Polyethylene is the same material as the ubiquitous plastic shopping bag, only much thicker. It has excellent light transmission and is highly diffusive, which is great for greenhouse growing. This is the cheapest type of greenhouse covering and has the lowest insulation value. We do not recommend single-layer poly.

Two layers of poly, however, can be used to create a bubble of air, which is an excellent insulator. This is called double-layer poly and consists of two layers that are sealed at the ends and inflated with a small fan. This is the least expensive option that we recommend.

The downsides to double-layer poly are:

- The walls can be easily punctured, though they are easy to patch.
- The average lifespan of poly in full sun is around 4–5 years, after which both layers must be replaced.
- Lower insulation value than polycarbonate.

Polycarbonate is a much harder material than polyethylene. For greenhouse coverings, it is made into corrugated panels similar in structure to cardboard. The corrugated structure not only gives it strength but provides many pockets of air in a single panel, which creates a higher insulation value than double-layer poly.

Polycarbonate panels are available in a variety of thicknesses, from 6mm to 16mm, and as twin-wall or triple-wall (two layers or three layers). The thicker the material and the more walls it has, the higher its insulation value. For a polycarbonate covering, we recommend a minimum 8mm twin-wall.

Advantages to polycarbonate walls:

- Superior insulation if using thicker than 8mm
- Very difficult to puncture
- Long lifespan (estimated 25 years)
- Looks fantastic

The only substantial downside, and it is significant, is that a new polycarbonate greenhouse will cost about three times as much as a double-layer poly greenhouse.

A final option is to use double-layer poly for the roof and polycarbonate for the endwalls. The endwalls contain a lot of the ins and outs of the structure, including cables, pipes and ducts, and it is very easy and neat to run them through polycarbonate (simply cut or drill a hole) and very challenging to run them through poly. This option allows you to keep costs lower while having excellent function, and it looks great.

Recommended Features

We strongly recommend you select all of these greenhouse features:

- Large doors on the west endwall. The capacity to easily bring in a tractor or other large machinery or equipment should be considered mandatory if at all possible. We suggest sliding doors that are at least 8′ wide.
- Roll-up sides. If you are selecting a poly greenhouse, this feature is highly recommended. The sides will roll up several feet via a simple crank and allow you the highest level of ongoing control over the ventilation of the space with no power usage. Ridge vents are another option and will vent heat more quickly but tend to be expensive and easily damaged.
- Automatic peak vents. This powered ventilation should be used in conjunction with the roll-up sides. It typically consists of mechanical louvers in one endwall and a large blower in the other that are controlled by a thermostat. They are standard in most greenhouses.
- Vented propane or natural gas heater. This is necessary in temperate or colder climates unless you utilize another heating method.
- Circulating fans. These are typically attached to the cross braces of the arches (the horizontal beams that look like rafters) and are used in the winter in conjunction with the heater to ensure that warm air is circulated within the greenhouse.

Heating the Air

In the winter when temperatures drop below the ideal average, supplemental heating is required to maintain a minimum air temperature of at least 5°C (41°F).

Vented propane greenhouse heater.

Air heating is best accomplished with a natural gas or propane heater. Your greenhouse manufacturer will be able to present you with the correct option that is sized for your greenhouse and climate. The heater, which is usually hung from the cross braces, is set to keep the air at a minimum temperature of 5–7°C (41–45°F). Circulation fans, also hung from the cross braces, are spaced around the greenhouse in order to rotate and mix air, ensuring air temperature is consistent.

Propane use will vary greatly from winter to winter. Most years our total annual propane bill is around $600, whereas in 2016/17, when drafting this book, we used more than double this due to a severe cold snap (and the winter isn't over). Propane will also vary in price over time.

Cooling the Air

In summer, or whenever the air temperature is consistently above 25°C (77°F), it is necessary to remove excess heat from the greenhouse. The circulation fans are turned off, and the air is allowed to naturally stratify, with hotter air rising to the roof peak and away from the crops and the cooler air sinking to the floor. Options for cooling the air in your greenhouse include direct methods, such as electrically driven swamp coolers or high-pressure water misters; and passive methods, such as ventilation and shade cloth. Our preference is for passive cooling methods, as they do not use any electricity, in combination with active ventilation via peak vents in the endwalls.

The primary method of passive cooling is to physically open the greenhouse to allow wind to pass through. This is typically done via either roll-up sides or ridge vents. Ridge vents are panels located at the ridge of the roof, running the length of the greenhouse. When opened, they allow the hot air to escape directly from where it is accumulating — at the top of the greenhouse. Although very effective, ridge vents can be easily damaged by wind and snow, and can be expensive to repair.

Roll-up sides are common on poly greenhouses and work just like they sound: a tube with a crank on one end is used to roll up the double-layer poly from the baseboard to about six feet high. This method is slightly less effective at venting the heat than ridge vents, as they vent the bottom instead of the top of the greenhouse, but they are more effective at venting humidity, which can be very useful in the winter to rapidly remove excess humidity.

The other passive cooling tool we recommend is shade cloth, which is used to cut down on solar radiation (direct sun exposure) and thus heat (see Chapter 5).

The last method of cooling we recommend is powered ventilation, which is simply using a fan to actively push hot air out of the greenhouse. Vents are installed in the peaks of the endwalls and are controlled by a thermostat. One endwall will have one or more large fans to blow hot air out and the opposite endwall will have an electrically controlled vent that opens at the same time to allow air into the greenhouse.

Roll-up side with crank.

Heating and Cooling the Water

Heating air takes considerably more energy than heating or cooling water. We therefore recommend your system contain two separate heating units: one for air as described above and one for water.

Water temperature is one of the most critical factors for the well-being of an aquaponic system. We cannot stress this point enough. Water temperature regulates the metabolic rate, and therefore the growth, of the fish in the system; it dictates the fecundity of the beneficial bacteria; it determines the oxygen carrying capacity of the water; and it plays a role in the transpiration process of the plants. For these reasons, it is very important to keep the water temperature in the correct range and as stable as possible.

One of the advantages of DWC aquaponic systems is the thermal stability afforded by the large volumes of water present in the system. In addition to the thermal mass of water, extra insulation is provided by the floating rafts and by the earth that the tanks, troughs and sump are resting on. Additional stability can be added by wrapping the fish tanks in insulation.

Peak vent and double-layer poly inflation blower.

Insulated fish tanks.

Despite its thermal, stability, the water temperature will be affected by fluctuating air temperatures, and regular adjustment, up or down, will be needed. Here in the Cowichan Valley in BC, average summer temperatures can be 30°C (86°F) for weeks on end, and winters can fall below freezing. Since we began recording daily temperatures, we have seen the thermometer get as high as 36°C (86°F) and as low as -15°C (5°F) for short periods of time. The goal is to keep the air temperature higher than 5°C (41°F) and ideally lower than 25°C (77°F) and the water temperature at a consistent 15–17°C (59–63°F).

The system for water temperature regulation must be automated for both heating and cooling.

Initially we did not include a water heating/cooling system in our design. The results were unhappy fish and weak plants. We began exploring the various options available with an eye to using as little electricity or fuel as possible.

We had an Energy Assessment done by the Province of BC, which evaluated the primary options: ground source heat pump, air source heat pump, gas boiler and refrigeration system, and solar panels and refrigeration. Solar panels were deemed to be the most efficient, as they produce much of the energy onsite but were ruled out due to the high initial capital cost. An air source heat pump had the second-lowest annual operational costs and was nearly the least expensive capital cost to install, so we opted to use an air source heat pump unit designed for a swimming pool. The unit sits just outside the greenhouse and has kept our water at the ideal temperature since installation. We recommend using such a unit.

Heat Pump

The heat pump should be an air-to-water heat exchanger designed for aquaculture or a large swimming pool, capable of heating and cooling water as needed. Correct sizing is very complex and must be done by an HVAC professional. Your unit must be able to keep system water at 15–17°C (59–63°F) at all times throughout the year and have a defrost cycle that enables it to operate at sub-zero temperatures. The internal heat exchanger cannot be copper as this is not fish safe; it must be titanium.

Our heat pump.

Our heat pump is an Orion series reversible pool heater made by HVAC Concepts, which has 65,000 BTUs of heating capacity, 38,000 BTUs of cooling capacity and draws 3,500 watts of power when running. These figures are for reference only. Our unit is sized for our water volume (which is less volume than the design in this book), our local climate conditions and the thermal gains/losses of our greenhouse.

The Raincoast Aquaponics Greenhouse

Our greenhouse is an 80′×36′ structure with a double-layer poly covering and twin-wall polycarbonate endwalls. It has all of our recommended features: large doors on the west endwall, roll-up sides, automatic peak ventilation and circulating fans. We use a propane heater for air heating and a heat pump designed for a swimming pool to both heat and cool the water. The greenhouse was engineered for our specific location for both snow load and wind. We purchased it new.

We recommend the book *The Year-Round Solar Greenhouse* by Schiller and Plinke (New Society Publishers, 2016). It is not specific to aquaponics but is a useful guide for all things greenhouse.

Greenhouse Layout

The best way to think about an aquaponic setup is to follow the water from its lowest point to its highest and back to the lowest.

The lowest point in an aquaponic system is the sump, which is fully underground. The sump acts as reservoir and a mixing and water conditioning vessel. Temperature and pH are also controlled in the sump. From the sump, the water is pumped through UV sterilizers to the highest points in the

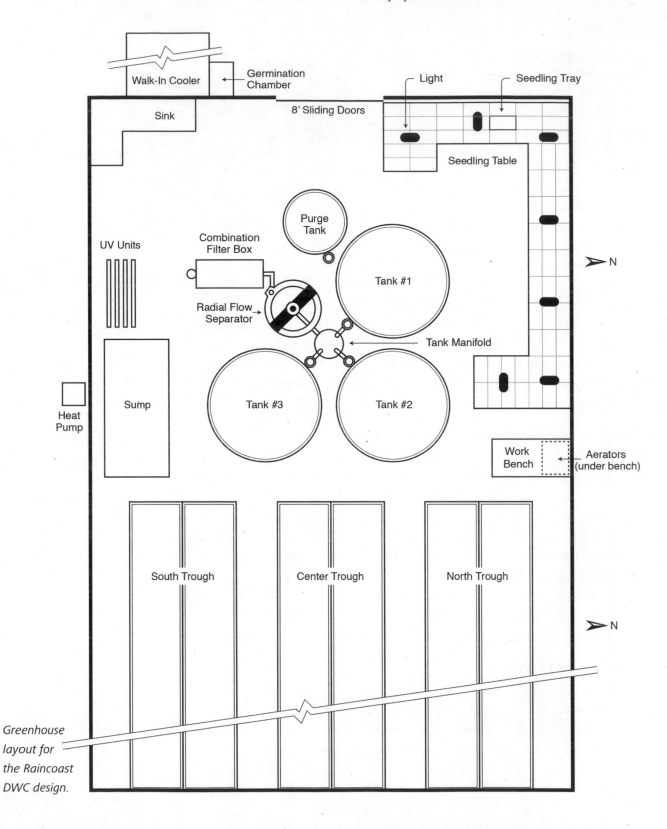

Greenhouse layout for the Raincoast DWC design.

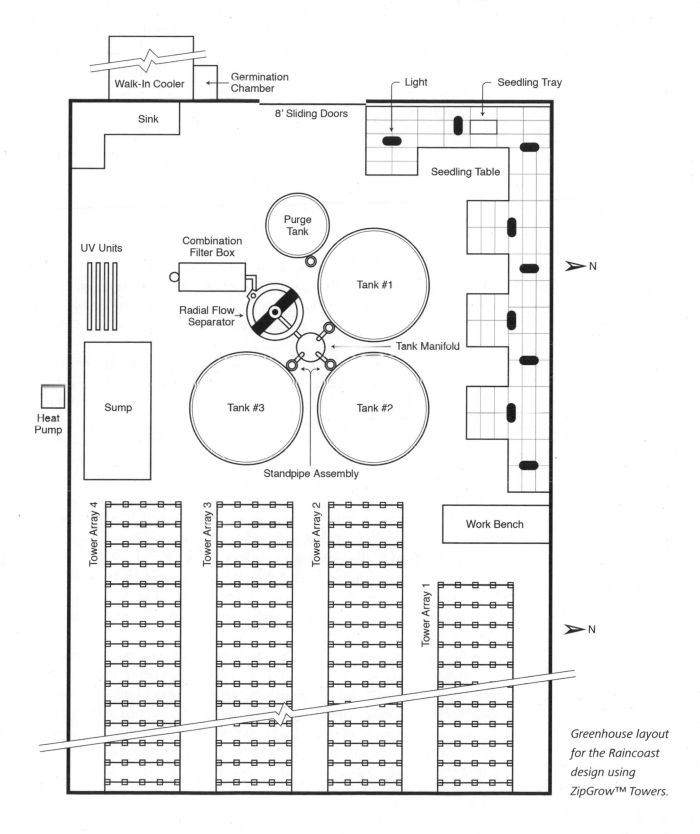

Greenhouse layout
for the Raincoast
design using
ZipGrow™ Towers.

system: the fish tanks. From the fish tanks, the water flows by gravity through the rest of the system and back into the sump. The water from the tanks exits through the Tank Manifold, then moves through the filtration systems (the Radial Flow Separator and Combination Filter Box) before flowing through the troughs and back to the sump.

In the RCA design, each trough has two sides. The water flows down the north side of each trough, moves to the other side via a U-shaped pipe at its eastern end, then flows back down the south side. A standpipe at the western end of the south side of each trough controls the water level in the trough. As the water overflows into the standpipe, it is carried back to the sump.

In a tower system, the sump is also the lowest point that acts as a mixing and conditioning vessel, but instead of pumping the water in one linear loop as in a DWC system, most tower designs use a split system. A split system uses two parallel loops, both originating in and returning to the sump, driven by separate pumps. The first loop pumps water from the sump through the UV sterilizers to the fish tanks, and from there it flows via gravity through the Tank Manifold and filtration systems back to the sump.

The second loop pumps water from the sump and distributes it to the top of the towers, which are arranged into multiple zones, or arrays. The water drips through the towers and collects in gutters under the towers. The water flows down the gutters and returns back to the sump. In a tower system, it is particularly important that the sump is designed for proper mixing of the two loops.

Troughs

Trough Design Principles

Space in a greenhouse is finite, and maximizing its use will directly equate to increased profits. We have designed our layout to maximize the usable production area and to allow for ease of work in each area.

A primary consideration for troughs is the commonly available dimension of styrofoam that will dictate the size of the rafts and thus the width of the troughs and trough layout. The trough layout is what determines that the greenhouse be at least 36′ wide.

The common dimension of styrofoam, including what we use and recommend, is 8′ long × 2′ wide. This is far too big to comfortably handle, especially when fully loaded with lettuce and wet roots. A loaded raft of this size could easily weigh 70–80 lbs and would break if you tried to lift it.

The solution is to cut the sheets in half into 4′ lengths that can be easily managed. This 4′ length dictates the width of one side of a trough and is the dimensional starting point from which to begin designing the overall layout of the greenhouse.

In the 120′ greenhouse design presented in this book, 86′ is dedicated to the troughs. Each side of a trough will hold 43 rafts. If you opt for a differently sized greenhouse or rafts, remember that the raft dimension dictates the inside dimension of the trough. You must also allocate enough space for the thickness of the three walls of the trough frames plus an inch or so extra per side to allow for rafts to be easily inserted, moved and removed.

The second major consideration is the layout and spacing of the holes in the rafts, which determines how many plants can be grown and will directly impact plant health and growth. If plants are too far apart, you will be wasting growing space and reducing yields. Plants too close together will lead to stretching, poor growth and increased disease and pests.

To complicate things further, consider that plants will be quite small when entering the rafts and quite large when harvested. Inserting young plants into the rafts with the more widely spaced layout required for mature plants will not maximize available growing space. Alternately, inserting plants in a tight pattern but not thinning or spacing them as they mature will lead to crowding. Neither of these is satisfactory.

Meeting the demands of growing plants by using rafts with different hole layouts requires buying lots of expensive styrofoam and storing it when not being used. Our solution is to use rafts that have a hole layout that meets the needs of both young and mature plants.

Each 4′×2′ raft has 32 sites, evenly spaced in rows 4 wide by 8 long. Each site is 6″ on center from its four adjoining neighbors and will allow you to use one raft design for all stages of plant growth. Immature plants are inserted into the rafts using all 32 sites. When the plants start touching each other (about two weeks in), they are double-spaced in a diagonal pattern until harvest. See Chapter 9.

The RCA Troughs

The tanks and work area will require 30′ of greenhouse length. For our recommended greenhouse length of 120′, the plant production area will be 86′ with the remaining 4′ as a walkway at the east end.

One of the unique qualities of the RCA design is the two-sided trough, developed to maximize the available growing area. By using back-to-back sides rather than the commonly employed method of a 4′ trough with walkways on each side, one more 4′-wide trough (one side of a trough in our design) can be fit in the width of a 36′-wide greenhouse. This is an increase of 20% in total plant growing area. In addition to the extra production capacity, this design is cheaper to build as there is less wood required to build the trough frames and far less pipe required. The only downside is that the side of a raft farthest from the walkway can be difficult to reach for shorter people. This can be overcome with a little bit of "greenhouse yoga" or by simply lifting one end of the raft out and pulling it closer to the walkway.

The total width of each trough is 8′9″: 8′ for the rafts, 3″ for the gaps between the rafts and frames (1.5″ per side), and 6″ for the thickness of the three trough walls. In a 36′-wide greenhouse, this will leave 29″-wide walkways. A 40′-wide greenhouse will have 41″ walkways.

The trough frames are constructed of lumber and plywood lined with Low Density Polyethylene (LDPE). The LDPE must be at least 20 mil, and no more than 40 mil. Note that "mil" stands for thousandths of an inch, not for millimeters. We strongly recommend using 20 mil, as it is thick enough for the job and much more pliable than 40 mil. The LDPE must be white on at least one side (the side that you can see when installed). The color on the other side is irrelevant. You must use LDPE as it is a food-grade contact surface and totally non-toxic to all organisms. Vinyl liner, although cheaper, has a tendency to leach plastic softeners into water over time, and many vinyl pool liners have anti-microbial chemicals manufactured into the plastic to control algae and bacterial growth. These chemicals are helpful to keeping a pool clean but very detrimental to an aquaponic ecosystem. There are also EPDM liners designed for ponds that are advertised as "fish safe" but are not a food-grade contact surface. Use only LDPE.

Your troughs will be 13″ tall. The water in the troughs will be approximately 10″ deep.

It is important that the surface on which the troughs are resting is perfectly flat and well compacted. After the trough frames have been assembled, a 1″-thick bed of masonry sand or other clean, rock-free sand is spread inside the frames and tamped flat. This is to ensure that the liner will not be punctured by any sharp edges, which is surprisingly easy when there are around 60 lbs of pressure per square foot pushing down on the liner.

Rafts

Rafts are made of a specific type of styrofoam: extruded polystyrene. This is the only type of styrofoam that is suitable for a DWC system, as it is chemically inert and will not leach anything toxic into the water. You must avoid all other types of styrofoam, notably anything with polyisocyanate, which is extremely toxic, or anything with a fungicidal or anti-microbial coating. If you have any uncertainty about the styrofoam, refer to the MSDS of the product.

Styrofoam typically comes in 2″-thick sheets of 8′×2′, which you will cut in half to make 4′×2′ rafts. Use "square edge" styrofoam. Avoid lap, pre-scored or tongue-and-groove styrofoam. Rafts thinner than 2″ should not be used.

Each raft will have a pattern of holes cut out of it, and be painted with 100% acrylic latex paint on one side for increased UV protection. 2″ net pots are seated in the holes and house the plants. Roots dangle through the styrofoam holes into the nutrient-rich water.

Fish Tanks

The basic requirements for a fish tank are very simple: a round watertight tank that has one large outlet at the bottom for water to exit. Water is input either over the lip of the tank or via a built-in inlet near the top of the tank.

In small aquaponic systems, a variety of reservoirs can serve as a fish tank. For backyard systems, a simple 50–60 gallon barrel can work well. For larger home systems or very small commercial systems, adequate tanks can be constructed from cisterns or can be made yourself.

For larger commercial systems like that described in this book, where having proper large plumbing connections is required and where a problem with a tank such as a sudden leak can lead to losses of large cohorts of fish, fiberglass tanks built specifically for aquaculture are the only sensible option. New fiberglass tanks for our design should cost less than $10,000 total and are worth every penny. We recommend against any type of home-made tank for a commercial farm.

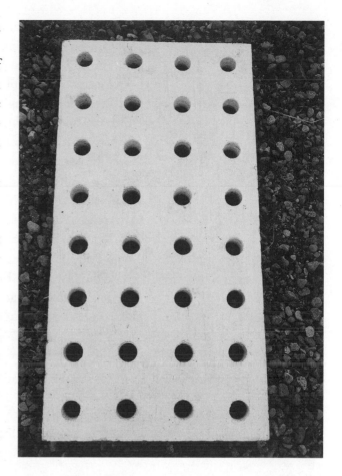

Raft shown with 32-hole spacing.

Aquaculture tanks are specifically designed and built to raise fish. Most importantly, they should never leak as long as they are properly plumbed, and they have a lifespan of about 40 years. They have a built-in large drain at the bottom that is essential for our design.

Good-quality fiberglass tanks are built to last, so you may be able to buy them secondhand. If possible, have secondhand tanks inspected by someone experienced prior to buying. In most cases, we strongly suggest purchasing new fiberglass tanks from a reputable aquaculture manufacturer. Our tanks were manufactured by PR Aqua and can be purchased through Pentair.

We use a four-tank system, meaning you will have four fish tanks. Three tanks are 8′ diameter by 3′ deep and are used to raise the cohorts. These are labelled Tanks 1, 2 & 3. The fourth tank is 4′ by 3′ and is used to purge the fish prior to harvesting. This is the Purge Tank.

The tanks are laid out in a specific pattern for optimum space usage, ease of work and ease of plumbing. The layout and installation is covered in detail in Chapter 4. Raising fish is covered in Chapter 8.

Each of the four tanks has an external Standpipe Assembly (SPA). Each SPA is connected directly to the bottom drain of a tank, controls the water level in that tank and acts as a raised drain by letting water exit the tank higher than the filtration components. The SPA also allows the tank to be drained, as needed, into the Waste Tanks. Construction and installation of the SPAs is covered in Chapter 4.

The final component in the tank system is the Tank Manifold. The Tank Manifold sits between Tanks 1, 2 & 3 and serves to combine the outlet water from the three tanks before it flows downstream to the filtration systems. The Tank Manifold is made from a 50–60 gal. food-grade plastic barrel.

The fish tanks, Tank Manifold and filtration systems (RFS and CFB) at our farm.

Filtration Systems

Filtration in an aquaponic system can be divided into two categories: mechanical filtration which removes solid waste and biological filtration which detoxifies ammonia and mineralizes particulates.

Filtration is one of the most critical components of an aquaponic system. It is often either under- or over-designed in its ability to effectively remove solids, process fish waste and control pathogens. There are numerous types and methods of filtration available from many different manufacturers, and there are an equal number of opinions on the best components and methods. The filtration systems we use are based primarily on commercial aquaculture standards and are proven to work in our design.

The primary design considerations when choosing a filtration system are:

- **Solids removal capacity.** Fish produce a surprising amount of waste every day. It is essential to remove all, or almost all, of the large particles, while allowing some of the fine suspended particles to remain in the system to be mineralized. Finer screens or mesh will capture more particles and require more frequent cleaning than coarse screens.
- **Water flow.** Water flow through the filtration components is powered by gravity. Accordingly, the filters must be able to capture solids without impeding or restricting the water flow as this could cause upstream components to overflow. Finer screens or mesh impede water flow more than coarse ones.
- **Cost to build or purchase.** This is self-explanatory: less expensive is better, as long as effective filtration is not compromised.
- **Energy consumption.** Powered filtration systems consume energy and increase your operating expenses. Minimizing power consumption saves money and makes your system more sustainable.
- **Ease and frequency of cleaning.** A good filtration system should be easy to clean and maintain. A poorly designed or undersized filter requires more frequent cleaning to ensure correct operation, which increases labor and water use. The difference between a system that requires cleaning every three days versus once or even twice daily cannot be overstated.

Mechanical Filtration

Mechanical filters will remove the largest unwanted items from your water including fish feces, excess fish feed, clumps of algae and other debris. The mechanical filters must be capable of trapping large amounts of solids without impeding or restricting flow of water, ideally without using any electricity.

There are a few different types of filters that fit these requirements, each with benefits and drawbacks. Some types of solids filters, such as rotary drum screens, require energy to operate and use large quantities of fresh water for backwashing. Other types, such as cartridge or bag filters, have no moving parts and don't require any energy to operate, but the water must be forced through them under high pressure, which requires extra pumps and thus more electricity. Neither of these is ideal for an aquaponic system.

Swirl filters and Radial Flow Separators have no moving parts, consume no electrical energy and allow a very high flow rate. Radial Flow Separators are a more modern design and are more effective. Our system uses an RFS.

The downside to an RFS is that while it is very effective at capturing large settleable solids, it does not capture smaller suspended particles. RFSs are also fairly expensive, although worth every penny in our opinion. Homemade versions tend to not perform nearly as well as commercial aquaculture versions. You can expect to pay about $3,000 for a new 36″ RFS.

Screen filters come in all shapes and sizes and can be easily made for or by you. In Chapter 4 we provide you with our custom design, sized specifically for our system.

Screens come in a wide variety of materials, and there are many manufactured options. By using the right filter body dimensions and screen types, these kinds of filters provide excellent solids removal capacity, good flow rates, low cleaning frequency and high ease of cleaning.

Biological Filtration

Biological filtration is the decomposition of organic materials by microorganisms. In an aquaponic system, this occurs in two distinct processes: nitrification and mineralization. We discuss these in Chapter 6.

Biofilters, like mechanical filters, can be bought or made and are available in a wide variety of types and sizes, from fluidized sand-bed filters and propeller-washed bead filters to trickling sponge filters and Moving Bed Bio-Reactors. Regardless of the type of biofilter, the principle is always the same: some form of media with a high surface area is contained in a vessel and water is passed through the media so that the bacterial colonies on its surfaces come into contact with the ammonia in the water, allowing the bacteria to oxidize the ammonia.

In DWC aquaponic systems, biological filtration occurs on the walls and floors of the troughs, the underside of the rafts and any other dark, well-aerated surfaces such as the inside of pipes. Plant roots also contribute to biological surface area and can roughly double the available BSA. In tower-based systems, the tower media provides the surface area for nitrifying bacteria to grow on. The amount of surface area in an aquaponic system that can be colonized by nitrifying bacteria is called the Bacterial Surface Area (BSA).

BSA directly correlates to the potential quantity of beneficial bacteria in your system and thus the processing power to convert ammonia into useful minerals. You can never have too much BSA. As discussed in Chapter 1, DWC systems have a relatively low BSA, so we designed our system to include a separate biofilter which is contained in the Combination Filter Box (CFB).

RCA Filtration Systems

Our system employs four stages of filtration. The first filters in the system are the fish tanks themselves. The circular design and the angle of the water as it enters the tanks create a constant swirling motion which forces waste to the bottom of the tank. By lifting the Standpipe Assembly (SPA) on each tank at least once per day, the majority of the collected effluent is removed from the system to the Waste Tanks. The second stage is a Radial Flow Separator (RFS) to trap large settleable solids as soon as they exit the fish tanks. The third stage is a set of screen filters which trap suspended solids. The fourth stage is a Moving Bed Bio-Reactor (MBBR). The screen filters and MBBR are both contained in the Combination Filter Box (CFB). The CFB greatly increases the BSA of the whole system.

Radial Flow Separator (RFS)

The RFS is a passive conical settling vessel with no moving parts that is designed to quickly and gently remove heavy solid waste from the system. It works in the same way as a swirl filter except that instead of moving the water around a horizontal axis, the water is moved around a vertical axis. Swirl filters were industry standard for some time. We recommend an RFS as it is about twice as effective at capturing solids.

Top of the RFS, showing the stilling chamber and the overflow weir.

Water enters the RFS in the central stilling zone and then is forced down into the settling chamber. As the water flows downwards, its velocity drops and its direction reverses into the outer part of the chamber, causing solids to quickly settle to the conical bottom. The accumulated solids are periodically flushed from the unit using the internal standpipe. Clean water exits the RFS by the overflow weir at the top of the unit.

RFSs are available from various aquaculture supply companies with some minor differences between them. We recommend a 36″ RFS with a 45-degree cone for a 120′ greenhouse using our system (note that 36″ is the diameter of the settling cone, not the actual diameter). An RFS of this size will be operating at slightly above its intended settling velocity but is recommended for our design due to space limitations. It will still be very effective at trapping large solids.

Combination Filter Box (CFB)

The CFB was designed by us and must be custom built by or for you. It houses filter screens of various porosities and a Moving Bed Bio-Reactor (MBBR).

The CFB is a rectangular box 2′ wide by 2′ high by 5′ long and can be made from food-grade plastic, aluminum, stainless steel or fiberglassed plywood. Unless you can find a prefabricated plastic box of the exact dimensions or are handy with welding and metal fabrication, we suggest building the CFB out of plywood and fiberglassing the inside to waterproof it.

The box has a water inlet on the upstream side facing the RFS and a water outlet on the opposite side which leads to the troughs. It has a removable lid and is fitted with a series of ribs on the inside walls that form slots for holding filtration screens in place.

The filter screens are made of curly plastic fibers woven into mesh sheets. The screens are arranged in a specific order, with the lowest density upstream

CFB with the lid removed. Note this photo shows the CFB on our farm, which is a slightly different design.

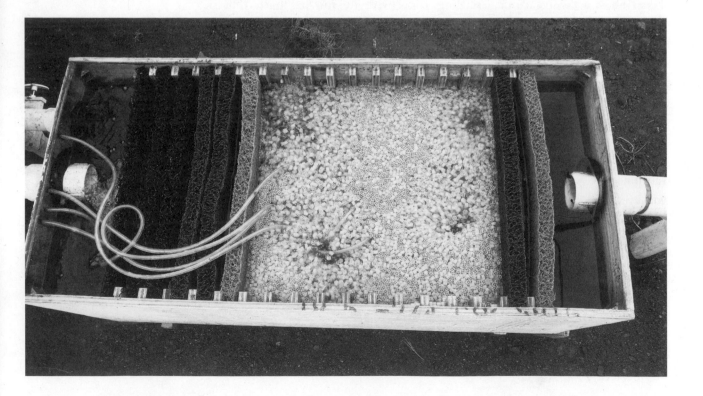

followed by mid density and fine density. The screens are removed and cleaned about every 3–4 days. We use and recommend screens manufactured by Matala.

The MBBR is a simple but effective type of biofilter. The open space between the filter screens is filled with a biological filter media, which provides a large surface area (BSA) for nitrifying bacteria. The area is vigorously agitated using air stones connected to the aeration system. The media we use and recommend is Sweetwater SWX bio-media.

The CFB is designed to exact dimensions for the design in this book. See Chapter 4 for construction details.

Ultraviolet Sterilization

Ultraviolet (UV) sterilization controls algae and harmful pathogens, and we consider it mandatory for both DWC and tower systems. The UV light eliminates organisms by rupturing their cell walls and destroying their DNA. Moving water through UV units at the required intensity will kill virtually all harmful and beneficial organisms.

As an aquaponic system relies heavily on beneficial bacteria, it is essential that the UV only be placed between the sump and the fish tanks. Using UV in this manner does not impact the beneficial bacteria cultures downstream and maximizes fish health by moving sterilized water to the fish tanks.

In addition to pathogen control, UV dissociates free chlorine and chloramine in the water, both of which are common in many municipal water sources and both of which are toxic to fish and bacteria. UV may also degrade certain botanical pesticides that could potentially make their way into the water. Pyrethrins, for example, is well-known to degrade quickly with exposure to sunlight and in the presence of water, so the use of UV may provide a safety net for some pesticide residues in your water.

UV light is classified as UV-A, UV-B or UV-C, corresponding to the wavelength in nanometers (nm) of light radiation produced. The peak germicidal efficiency of UV light is 264 nm, which is in the UV-C range (200–280 nm). Your UV units must be rated for UV-C output. Most UV-C lights emit at 254 nm which is very close to ideal.

There are three types of UV bulbs used in treatment systems: high, medium and low pressure. Medium-and-high pressure bulbs are polychromatic and emit UV light across a range of wavelengths. These bulbs are typically much more expensive, have a shorter lifespan than low pressure bulbs and are better suited to applications such as sewage treatment or swimming pools. Low-pressure UV bulbs are monochromatic and emit one wavelength of UV light, making them ideal for microbial sterilization in aquaculture and aquaponic systems. Low-pressure bulbs are also less expensive, more efficient and generally have a much longer lifespan (1–2 years).

IMPORTANT

The UV lights must be low-pressure UV-C lights at 254 nm.

The fluence, or UV dose, of an ultraviolet bulb is measured in millijoules per square centimeter (mJ/cm^2). The UV dose is directly impacted by the speed of the water as it moves past the bulb. The slower the water, the longer it is exposed to UV and thus a higher UV dose. The faster the water, the less time it is exposed and thus a lower UV dose.

For example, the units we use on our farm (Smart UV High-Output E150S) put out 180 mJ/cm^2 at 65 liters per minute (LPM), but only 30 mJ/cm^2 at 387 LPM. The water on our farm flows at approximately 130 LPM. We use two UV units in parallel, which means the total flow through each unit is 65 LPM. Accordingly, each unit is putting out a theoretical dose of 180 mJ/cm^2.

We say "theoretical," as the mJ/cm^2 rating you will see listed on all UV lights is based on testing done by the manufacturer under perfect conditions. The actual transmission of UV light through water, known as Ultraviolet Transmittance (UVT), will always be less in an aquaponic system due to water turbidity (mainly effluent and algae in the water) and the high dissolved mineral content.

While the quality and clarity of water will vary from farm to farm and farmer to farmer, the commonly accepted average loss of UVT in an aquaculture (or aquaponic) system is approximately 25%. This means that at our farm the theoretical output of 180 mJ/cm^2 is actually closer to 135 mJ/cm^2. In a recent UVT test by an independent lab, our water tested at 80% UVT (20% loss).

IMPORTANT

Adding humic acid to system water, commonly done to shade the water and prevent algae growth, must not be done in our design as this will dramatically decrease UVT.

The UV dose required to kill an organism will vary widely from species to species. Common algae, such as *Chlorella vulgaris*, are controlled at 22 mJ/cm², and the pathogenic *Aeromonas salmonicida* is controlled at a very low 3.6 mJ/cm². However, *Flavobacterium psychrophilum*, which we have found to be the most UV-resistant pathogen in our system, is only destroyed at a dose of 126 mJ/cm². We therefore recommend that the UV sterilizers be capable of giving a UV dose of at least 130 mJ/cm². Factoring in UVT losses of 25%, the units should be rated at a minimum of 180 mJ/cm² at your flow rate.

If you are using our design with three 8′ diameter fish tanks, with a minimum required water flow of 270 LPM, you would require four of the same UV units we use to achieve a theoretical UV dose of 180 mJ/cm² and an actual UV dose of approximately 130 mJ/cm². If your system moves water more rapidly, you will require additional or larger units to maintain the required UV dose. See Pumps later in this chapter.

UV bulbs should be replaced every year. While they will continue putting out light long past this time, the output is constantly degrading and will degrade to unacceptable levels that may not effectively control all pathogens. See Chapter 11 for replacement instructions.

Initially, we did not include UV sterilization in our system, and we had ongoing disease issues with our fish. We lost whole cohorts. While we cannot state with certainty that UV solved all disease issues, since installing UV sterilization, we have had only minimal disease issues and have lost no cohorts. UV is not a guaranteed solution to all disease and pathogens, but we strongly believe it greatly supports fish health as long as your water, as it enters the UV units, appears clear enough to drink. Once your system is established, we recommend testing the UVT of your water to ensure that the correct dose is being achieved.

Our UV units cost approximately $1,000 each, the replacement bulbs approximately $120, and each unit draws a continuous 150 watts of power. We have never regretted the expense. We consider adequate UV sterilization mandatory.

Supplemental Lighting

In order to produce year-round at most latitudes where cold-water systems are more prudent, supplementary light will be required for at least part of the year.

For optimal production, plants require a certain quantity of light per day. Light is often measured in lumens, but the superior measurement is PAR (Photosynthetically Active Radiation, which is measured in micromoles. A lumen is a measure of the total light visible to the human eye. A PAR micromole is a measure of the total number of light photons (particles) that is usable by plants. The problem with using lumens is that plants view and use light very differently than humans.

There are some 602 quadrillion photons in each micromole. A mole is one million micromoles. Each type of plant has its own light requirements for optimal production. In general, lettuce will require about 250 micromoles per square meter per second, or about 17–25 total moles per square meter per day. This is called the Daily Light Integral (DLI), a measure of light intensity over time.

Gavita Pro 1000-DE on a LightRail mover.

Full sunlight is around 2,000 micromoles per square meter per second, which means that 25 moles is achieved in about 3.5 hours of full sunlight. However, as conditions are rarely optimum and delicate plants such as lettuce

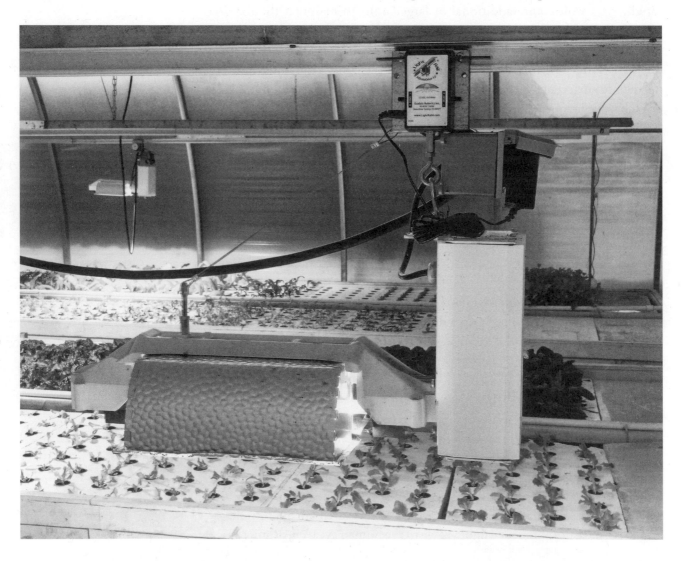

cannot handle full sunlight, this must be spread out throughout the day. In summer, shade cloth protects the lettuce from full sunlight, but in winter when light levels are poor, supplemental lighting is required to achieve a DLI of 17–25 moles per day.

As a general rule, lettuce needs no less than 14 hours of sunlight per day. According to the Cornell University CEA *Hydroponic Lettuce Handbook*, commercial lettuce production requires 16–18 hours per day. At our farm in southwest BC, we use supplementary lighting at various levels for about six months per year — from fall equinox to spring equinox. Of this period, months 1 and 6 generally are lit for two hours at either dawn **or** dusk; months 2 and 5 are lit for two hours at dawn **and** dusk; and months 3 and 4 (the heart of winter) are lit for 3 or more hours at dawn **and** dusk.

HID (High Intensity Discharge) lighting may make your greenhouse visible for miles in the dark, and your neighbors may not appreciate this late at night or early in the morning. Accordingly, we suggest dividing your supplementary lighting to be on at dawn and dusk.

We recommend using 1,000W HID lamps on light moving rails. A 1000W light hung 4 feet above the plants will cover the width of a trough (8′) and when installed on a light mover can cover about 20 linear feet of the trough. For a 120′ greenhouse with 86′ troughs, you will need 4 lights per trough, or 12 lights total for three troughs.

We use and highly recommend Gavita Pro 1000W DE lights for the troughs. We also use and recommend LightRail light movers. See Chapter 4 for installation instructions.

You will also require supplemental lighting for the germination chamber and seedling area.

Germination Chamber

Our system is designed to start each plant from seed in a germination chamber before moving the seedlings into the greenhouse. Alternatively, you can purchase plant starts. Achieving a high level of germination depends mostly on the environmental conditions, notably temperature, humidity and light, as well as the age and quality of the seeds.

This design incorporates a climate-controlled Germination Chamber attached to the walk-in cooler that is sized to produce sufficient seedling plants for one week of production. For three 86′ troughs, the box should be capable of holding at least 15 standard nursery trays that are 11″×22″.

If you are building a system of a different size, you can determine the amount of seedlings needed to germinate every week by dividing the number of plant sites in the troughs by the length of the expected growth cycle.

The formula to calculate the space requirement for seedlings is:

$$(A \times B \times C) / D \times 1.2 / E = X$$

Where A = number of trough sides
Where B = number of rafts per side
Where C = number of harvested sites per raft
Where D = average growth cycle of plants in troughs
Where E = number of cells per nursery tray
Where X = number of trays required

The calculation for our design is:

6 (trough sides) × 43 (rafts per side) × 24 (average harvestable sites per raft) = 6,192 plant sites

In an average growth cycle of 5 weeks, you can expect:

6,192 / 5 = 1,238 plants per week

plus 20% overage to allow for losses = 1,486

1,486 / 98 cells per tray = 15.16

We round this down to 15 trays. The Germination Chamber design is two levels and holds 18 trays total. See Chapter 4 for construction details.

The ideal temperature to germinate most seeds is 15–18°C (59–64°F). Heating will be accomplished via a small electric heater or heating mats under the trays. Cooling is accomplished by stealing some of the cold air from the walk-in cooler via a small fan through the adjoining wall. Both the fan and heater are controlled by a dual-temp thermostat that turns the heater on when too cold and the fan on when too warm. The chamber is insulated to minimize the impact of extreme outside temperatures.

The humidity of the seedling trays must be kept between 75% and 85%. This is accomplished by placing standard humidity domes over the seed trays for the duration of the germination period plus a few days in the seedling area.

Seeds do not like intense light. We recommend using a single T5 fluorescent lamp over each tray. Lamps should be 6,000K or higher. The lamps must be on for at least 20 hours per day and can be left on 24 hours per day. We suggest 20 hours per day.

The first leaves to appear are the cotyledon. These are the tongue-like embryonic leaves contained within the seed that allow young plants to absorb initial much-needed light. Cotyledon look different than true leaves. True

leaves grow between the cotyledon and signal that the seedlings are ready to move as soon as possible to an area of light intensity close or equal to what the mature plants are grown in.

Our Germination Chamber. Note that ours is only one level, and we use different lights.

Seedling Area

The seedling area is the intermediary, second stage of plant growth in the system. Here the young plants, with only their first set of true leaves, will grow to be sturdy small plants over 3–4 weeks. During this time, the plants will establish strong roots and grow subsequent sets of true leaves.

Keep the lids on the trays for the first 2–3 days after moving them from the Germination Chamber. For lettuce, after 3–4 weeks you can expect the plants to be about 2–3″ tall with 3–4 sets of leaves, at which time they are large enough to transplant into the main system.

Once the domes are removed, the environmental conditions for the plants will be those of the greenhouse. For best results in the cold season, we strongly suggest adding both under root heating and supplemental lighting. The seedlings will only require bottom heat for the first couple of weeks after they are moved from the Germination Chamber.

Heating mats are the simplest and most effective root heating option. Available in a variety of sizes to fit your tables, the mats must be 100% waterproof. We recommend heating mats.

Seedlings will require supplemental lighting for the same period as the older plants in the troughs. Of all plant stages, seedlings require the highest amount of daily light. Accordingly, while the older plants in the troughs will do fine with lights on light movers, thus less exposure, the seedling area should be illuminated in totality when the lights are on. For this purpose, we suggest using 400W Metal Halide (MH) or enhanced High Pressure Sodium (HPS) lamps. You can also use 600W lamps, which can cover a wider area.

Mount fixed lamps over the seedling area tables. The lamps should be kept approximately 2′–3′ above the plants at all times and will cover up to 6′ of table length per lamp. See Chapter 4 for mounting options.

There are a number of methods of watering seedlings including several automated methods. We recommend simple top watering by hand using a small pump and hose with a diffusive nozzle. The water from the troughs is the ideal source. Simply place an oil-less pump into the outlet end of a trough.

The plants require water once to twice daily, depending on their maturity and the air temperature. Watering takes just a few minutes and is a good opportunity to regularly observe the health of your plants. Remember that the younger the plant, the more delicate it is. Regular observation and adjustments pays dividends. Remove dead or weak plants immediately.

We strongly recommend that your Seedling Table be only one level rather than stacked. The obvious disadvantage to this is the area required. We experimented with numerous stacked designs and artificial lighting options and found all unsuccessful or not nearly as successful as a single level. Problems with multi-level tables include poor growth due to a lack of natural light, humidity issues and increased pests. Having the seedlings unshaded in the open greenhouse environment far outweighs the extra space required.

The total area required for the Seedling Table is calculated by multiplying the number of seedling trays germinated each week by the number of weeks they will spend on the Seedling Table. We know that for a five-week cycle (in the troughs) we require 15 trays to be germinated per week. Plants will spend 3–4 weeks in the seedling area, thus the Seedling Table must be able to hold at least 60 trays (15 trays × 4 weeks). The layout as shown in the diagram DWC Overview on page 30 allows for 63 total trays.

Water: The Lifeblood of the Farm

One of the biggest advantages of aquaponics, both for the farmer and the environment, is the dramatically reduced water consumption over traditional farming. On a traditional soil-based farm, the average head of lettuce uses up to 15 liters of water during production. We use less than 2 liters to produce the same lettuce. Once you are operational, you can very accurately calculate the total water per plant by metering the total water added to the system for a week then dividing this amount by the number of plants harvested. Calculating this over longer time periods will increase the accuracy.

Despite the small amount of water "consumed" by the system, large amounts of water are present in an aquaponic farm. It is constantly swirling through the fish tanks and flowing down the troughs (or trickling down the towers). Our 120′ DWC system will have approximately 66 cubic meters of water in it at all times. A 120′ drip tower system will have approximately 20 cubic meters.

The majority of water used by the system is due to flushing the fish tank SPAs and the RFS. Flushing occurs when you quickly remove and replace (or "pop") the SPAs on the tanks and the RFS for less than a second, thus flushing collected large particulate matter from the system. Transpiration and evaporation account for the additional water usage.

Monitoring and regulating water parameters is one of the most important roles of an aquaponic farmer. Often this will simply mean ensuring that the electronic monitors and controllers are functioning properly, but it will also mean conducting periodic manual tests that are more foolproof than automated tools. The water quality parameters of note are: temperature, pH, ammonia, nitrites, nitrates and mineral content (notably calcium, potassium, magnesium and iron).

Water Temperature

As discussed in Chapter 1, there are several reasons and advantages to farming in a cold-water system. The optimum temperature for your system is 15–17°C (59–63°F). You should strive to keep your water within this range at all times and no more than plus or minus 2 degrees. This should not be a problem as the temperature will be controlled by a heat pump.

Between 15–17°C is the range in which the balance of the different biological needs in your system is best aligned. We operate our system at 16°C (+/- 1°), which is ideal for fish and within the ideal range for plant growth. It is, however, very low for the bacteria, which have a 50% decrease in ammonia oxidizing capacity compared to 25°C (77°F) water. Even a 1–2° increase in temp above 15° will significantly improve bacterial performance, but this comes at the expense of trout health because that species is not well-adapted to warmer temps.

Aquaponics requires electricity to run, and our goal is to design and operate the system as efficiently as possible. Aside from the supplemental lights, conditioning the water is the biggest electrical consumer in the system. Maintaining the water temp at or near to your average local annual temperature will greatly minimize electrical consumption. This means that cold-water aquaponics is less suited to tropical or warmer climates and is ideally suited to temperate climates or colder.

IMPORTANT

The ideal aquaponic system raises a fish species that can thrive at a water temperature near the local annual average, grows varieties of plant species well adapted to the local growing conditions and maintains a water temperature and pH that is high enough to support good bacterial performance.

pH

pH (potential of hydrogen) is the acidity or alkalinity of a substance; 7.0 is neutral. Less than 7.0 is acidic, greater than 7.0 is alkaline. pH is critical to the health and performance of an aquaponic system. It is essential you understand the processes involved in managing pH effectively.

Managing pH is another balancing act between the different biological needs in the system. The ideal pH is one that heavily favors the needs of the plants (generally below 7.0) without being too low for the fish and bacteria. Fish and bacteria prefer to be in water above 7.0 but are generally tolerant of mildly acidic water. The reason plant growth is heavily weighted in this consideration is that the plants generate the vast majority of the revenue, thus their rapid and vigorous growth without excessive detriment to fish or bacteria will greatly increase revenue.

We have found the ideal pH to be 6.4 to 6.6, and we try to consistently maintain a pH of 6.5.

pH for the system is always measured in the sump near the pump inlets. The goal is to measure the pH just before the water enters the fish tanks. A pH controller that contains both a pH monitor and a dosing pump is located beside the sump and automatically adjusts the pH as needed.

The processes of nitrification and mineralization tend to constantly make water more acidic (lowering the pH). In most cases, you will need to compensate for this by slowly adding an alkaline chemical to maintain the pH at the desired setpoint. However, in some locations, the municipal or well water may have a high mineral content (known as "hard" water) due to it passing through limestone formations in the aquifer. The carbonates present in hard water will tend to buffer the pH of the aquaponics system, keeping the pH fairly high (above 7.0). If the level of carbonates in the incoming water is high enough, you may find that you need to continuously adjust the pH down.

At our farm, the source water is very "soft" (low alkalinity), so we only need to adjust the pH up. pH is adjusted up by introducing potassium hydroxide (KOH or caustic potash) and calcium hydroxide ($Ca(OH)_2$ or hydrated lime) through two separate means.

KOH, which is readily available in a concentrated liquid, is dosed into the system via the pH controller. $Ca(OH)_2$, which is most commonly available as a fine powder, is added to the system by putting a sock full of it in the Combination Filter Box to slowly leach out. This method, though seemingly inelegant, is highly effective. Never use sodium bicarbonate or sodium hydroxide in an aquaponic system, as the sodium can be toxic to the plants.

KOH and $Ca(OH)_2$ are used to adjust pH because, at the levels needed, they are non-toxic to the fish, plants and bacteria and have the added benefit of supplementing calcium and potassium, important plant nutrients that tend to be deficient in most aquaponic systems.

If your water supply is particularly hard and you need to adjust the pH down, use phosphoric acid (H_3PO_4). As is the case with hydroxides, at the levels needed, phosphoric acid is non-toxic to the fish, plants and bacteria and has the added benefit of supplementing phosphorus to the plants. In this case, you may need to also supplement small amounts of potassium and calcium, depending on the levels of these minerals in your incoming water. Phosphoric acid is dosed into the system by the pH controller.

Water Quality Management

Water quality tests should be performed daily and weekly in order to monitor the health of the system and make adjustments. It is critical that ammonia and nitrite levels be tested frequently as this is the only way to monitor the ongoing performance of the biofilter. Minerals that are important for plant growth should be tested at least weekly to make sure that they are present in sufficient quantity.

See Chapter 6 for a detailed discussion on pH and water management.

Aeration

All the beneficial life in an aquaponic system requires oxygen. Fish "breathe" it, plants absorb it through their roots, and bacteria consume it in their metabolic processes.

In the root zone of DWC systems, plant roots are fully submerged in water so the only oxygen available to them is dissolved in the water and is limited to about 10 ppm at 15°C, and even less at higher temperatures. The availability of oxygen at the root zone is one of the limiting factors in a DWC system and may be the reason why certain plants, such as spinach, do not thrive in troughs but do well in towers. It is best to think of the root zone in a DWC system as **air saturated water**.

In ZipGrow™ Tower systems, plant roots are held in a very porous media that has a void ratio (or open space) of about 90% through which the water is trickled. Because the roots are not fully submerged as in a DWC system, the roots are able to absorb oxygen from both the water and the ambient air which contains far more oxygen. It is best to think of the root zone in tower systems as **water saturated air**.

As the water falls through the tower media, it is broken into many fine droplets with a high overall surface area which enables oxygen in the air to dissolve into the water. This makes the towers self-aerating and negates the need for a mechanical aeration system.

In DWC systems, there is no natural mechanism to create a large contact zone between water and ambient air which would allow sufficient oxygen to dissolve into the water. Without supplemental aeration, all oxygen already dissolved in the water would quickly be used up by the biological activity of the bacteria, plants and fish, and the entire system would suffocate in a matter of hours.

Supplemental aeration is accomplished via air pumps called regenerative blowers (we call them "aerators" throughout this book) that force ambient air through underwater diffusers called air stones.

Air stones are porous silica and break up air into small bubbles. Smaller bubbles have a far greater collective surface area than larger bubbles. More surface area means more oxygen is able to dissolve into the water. Air stones are placed throughout the troughs, in the sump and in the CFB. They are never placed in the fish tanks as doing so impedes the self-cleaning nature of the tanks.

In the troughs, we recommend using one Sweetwater AS5 air stone under every raft. These stones are 3"×1" and have a suggested flow rate of 0.30 CFM each. For the sump and the MBBR, we recommend using Sweetwater AS15 air stones (ten in the sump, six in the MBBR). These stones are 6"×1.5" and have a suggested flow rate of 0.50 CFM. You will also require MNPT x Barb connectors and air tubing.

Aerating the troughs has the additional benefit of gently agitating the plant roots which prevents excessive particulate accumulation on the roots. Aerating the sump serves two purposes: to fully mix in the water returning from the heat pump and to homogenize the chemicals used for pH adjustments. Aerating the CFB vigorously agitates the MBBR and provides oxygen for the nitrifying bacteria.

One aerator of sufficient size can easily supplement enough oxygen for system saturation. The problem with having only one aerator is that there is no built-in redundancy. In other words, if your aerator fails (and like all mechanical equipment, it eventually will), your fish will die in a few short hours as the oxygen is used but not replaced.

There are two options for providing the required redundancy: have a second backup aerator onsite and/or connect the backup oxygen system to the aerator via the monitoring system so that if the aerator fails, oxygen will be supplied to the fish.

If you opt to have a backup aerator, it can either simply be onsite, ready to be installed if the active unit fails, or the two units can be installed in parallel with a pressure switch and relay between so that the second unit activates automatically if the first fails. If you do not use an automatic standby aerator, or forgo a backup aerator entirely, you must use a monitoring system and a backup oxygen system to deliver oxygen to the fish tanks in case the aerator fails. See Monitoring System and Backup Oxygen in Chapter 5.

The only downside to having two aerators onsite is the extra up-front expense. In the long run, this is moot as you will need to replace units over time and their life spans will not be shortened by being onsite. We cannot over-stress the importance of maintaining aeration. We strongly suggest having two identical aerators onsite at all times, ideally installed in parallel so

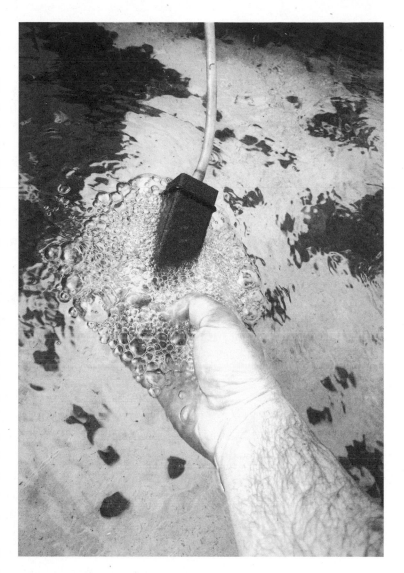

Air stone actively aerating the water. Notice the O-ring bumper.

that the backup activates automatically in the case of the primary aerator failing.

To determine the size of aerator needed for your system, add the flow rate of all air stones at their suggested flow rate.

If using our recommended quantity of specified stones:

Troughs:

$$258 @ AS5 \text{ stones} \times 0.30 \text{ CFM} = 77.4 \text{ CFM}$$

Sump and MBBR:

$$16 @ AS15 \text{ stones} \times 0.50 \text{ CFM} = 8 \text{ CFM}$$

$$Total = 85.4 \text{ CFM}$$

Therefore you will require an aerator capable of delivering a minimum of 86 CFM @ 10″ of water (the depth of troughs). For this example, we recommend the Sweetwater SST30 regenerative blower which outputs approximately 90 CFM at 10″ depth.

Pumps

Like the arteries in your body moving blood, the plumbing in your farm will move water throughout the system in a precise and controlled manner. The system is designed to move water within a specific range of flow at all times. Improperly designed systems can have water that moves too fast, which can cause poor ammonia oxidation and solids removal, or too slowly, which can create stagnation and low oxygen levels.

Water will be moved through your system via two parallel centrifugal pumps, though only one pump will be operating at any time. A centrifugal pump uses an impeller to move water. Most modern water pumps are oil-less, meaning they do not contain anything that is toxic if they physically break. Your pumps must be oil-less.

The key characteristic of a pump is its performance curve: a graph published by the manufacturer that shows the volume of water that the pump can move at a given head (resistance). The term "head" is used to define the amount of work a pump must do to overcome gravity, vertical height and friction in the pipe.

The "shut-off head" of a pump is the height at which a pump can no longer move water through the discharge pipe. You cannot operate a pump under this condition as it will burn out.

The "total head" is a cumulative total of: the height of the pump over the water level in the sump (called the suction head), the vertical distance between the pump and the water inlets to the fish tanks (called the vertical head) and the friction losses of each pipe and fitting between the pump and the fish tanks.

Friction losses are due to the friction of the water against the sides of the pipe; dynamic losses are due to turbulent flow created by the bends of individual fittings.

The objective of the pumps is to fully replace the water in each fish tank every 30–45 minutes. Each of the three 8′ diameter tanks contains approximately 4,000 liters (or 4 cubic meters) of water, so the pumps must be capable of moving approximately 12,000 liters of water every 30–45 minutes to replace all water in the tanks.

To properly size pumps for your system, you will first need to know the total head and the desired rate of flow. Once you know these two pieces of data, you can select pumps by looking at the pump curve published by the manufacturer.

Due to site specific conditions and farm preferences, each system will have a different total head. To discern the size of the main pumps required for your system, calculate the total head of your system once your plumbing is installed. As with the aerators, this design calls for two identical pumps, each sized to power your system by itself. The rationale is again redundancy, as all pumps eventually fail.

Early on in our system, we learned this lesson the hard way. Our original design contained only one large pump which worked well enough until the day that a wayward frog found its way into our sump, was sucked up into the impeller and caused the pump to seize. Before we could remedy the problem, we lost an entire cohort of fish. It was this frog incident that led us to prioritize pump redundancy and create the backup oxygen system (see Chapter 5).

The two pumps are installed in parallel with valves that allow easy switching from one to the other. The water flow output from the pumps will be monitored by the monitoring system which will inform you if the water flow drops below a set point. Optionally but recommended, you can install a flow switch and relay between the two pumps so that if the active pump fails, the other will automatically activate.

For our design, the minimum flow required is 270 LPM, which will replace the water in the three 8′ tanks approximately every 45 minutes. The maximum flow rate is 400 LPM which will replace the water every 30 minutes. We recommend a target flow rate of 400 LPM. Note that the flow rate can be too high. If it is much above 400 LPM, the solids retention capacity of the RFS will decrease, and higher flow requires greater UV capacity.

See Chapter 4 for details on calculating head and sizing pumps.

Tower System Pumps

If you are using a tower system, you will need one or more additional pumps to circulate water from the sump to the towers as part of a split system. A split system contains two separate loops rather than the single loop used in a DWC system. The first loop is mostly the same as in DWC: sump to UV, UV to tanks, tanks to Tank Manifold, Tank Manifold to CFB. From the CFB, rather than flowing to the troughs as in a DWC, the water returns to the sump completing the first loop.

The second loop runs from the sump to the towers and back to the sump again. Use the same process to calculate the total head for both loops and size your pumps based on their pump curves. In general, a 5′ tower requires about 20 liters per hour, or ⅓ of an LPM. For a 120′ greenhouse which can hold approximately 1,100 5′ towers, the pump on the second (tower) loop must be able to move 350–400 LPM.

Effluent

Fish poop. A lot. This is the main source of effluent (water containing waste) in an aquaponic system. Secondary contributors include uneaten fish feed, algae and plant debris. These wastes must be frequently removed from the system or it will rapidly overwhelm the nitrifying bacteria and create anaerobic zones, leading to high ammonia levels. Excess ammonia will quickly become fatal to fish.

Two bottle sizes of our bioactive fish fertilizer.

As discussed earlier in this chapter, our system employs four stages of filtration: the fish tanks, the RFS, and the filter screens and MBBR in the CFB. If using a tower system, you do not require an MBBR as the matrix media in each tower provides ample BSA.

The combined four stages of filtration will successfully remove enough solids from your system to maintain a healthy ecosystem. But what to do with the effluent? The simplest option is to spray the contents onto a lawn or field as a grass or hay fertilizer or to fertilize an orchard or garden.

We recommend fermenting, bottling and selling it as a local organic fertilizer. Properly produced, your "waste" product will be one of the best fertilizers in the world, highly prized by farmers and gardeners. We also recycle the finished fertilizer back into our system to provide additional minerals to our plants (see Chapter 6).

The fermentation process is actually quite simple:

1. Collect effluent in the Waste Tank. We suggest using IBC totes.
2. When the Waste Tank is full, transfer the effluent to a settling tank (another IBC tote). Leave it undisturbed for a day or two, allowing the solids to settle to the bottom.
3. The relatively clean water in the upper column of the settling tank is called the supernatant. The supernatant is still very rich in nitrogen and other plant nutrients as well as beneficial bacteria. Drain the supernatant from the settling tank via a valve about ⅓ of the distance from the bottom. It can be stored in another tank or used immediately for crop irrigation.
4. The concentrated solids are then transferred to 50–60 gal food-grade plastic barrels for biodigestion. We aim to have a concentration of 25–50% solids for biodigestion.

The biodigestion, or fermentation, process is also known as liquid composting. In this process, the bacteria contained in the effluent solution will break down the organic solids into their elemental constituents. These elements are plant food of the highest caliber. To ensure optimum fermentation, we recommend keeping the mixture at 20–25°C (68–77°F). We use one 300W aquarium heater controlled by a thermostat in each barrel, and we recommend wrapping the barrels in an insulative jacket to buffer temperature. You must also aerate the mixture by using an aquarium air pump with an airstone in each barrel. The air pump needs to put out at least 0.5 CFM at 36″ depth.

With good conditions, your effluent mixture will be converted into the highest quality fertilizer in 1–3 months. To gauge completion of the biodigestion process, test it for ammonia. Finished fertilizer will not have any ammonia in it. Once complete, we suggest sending it to a lab for a mineral analysis which can be included on a label and is useful for sales. We sell our fertilizer in 1L and 4L bottles at farmers markets and local garden centers.

We do not pasteurize our fertilizer; thus, it is still full of microbial activity. If not used within a few weeks, it may turn anaerobic and start to smell. This can be overcome by shaking the bottle and burping (opening) it. You can also include a use-by date on the bottle.

Alternatively, the fertilizer can be pasteurized prior to bottling to get a shelf-stable product. We choose not to do this as pasteurization is energy intensive and the product loses all the benefits of live microbial activity. We prefer to treat it as a "live" product and instruct our customers to use it within a few weeks or burp regularly.

The Sump

The sump is the lowest point in the system and is both the beginning and the end of the flow of water. Water is pumped out of the sump to the highest point in the system — the fish tanks — before flowing downstream through the system via gravity back to the sump.

The sump serves multiple functions, and its design and construction are just as important to the operation of the system as any other component. The sump acts as a reservoir for the collection of water returning from the hydroponic subsystem. It also serves as a mixing and conditioning vessel, a water quality monitoring station and a pump station.

The volume of the sump is important. First, it must hold enough water during normal operations to effectively mix and temper the return water from the troughs and the return water from the heat pump. Second, it needs to have enough capacity to handle the "drain down effect" in the event of a power outage. See Drain Down Effect.

In our design, the sump is a 4′ wide by 10′ long pit that is 4′ deep. It has reinforced walls and is lined with the same LDPE that is used to line the troughs. A 2×8 collar around the perimeter, sitting on top of the reinforced walls, allows most of the plumbing, air and electrical lines into and out of the sump. The collar also holds the sump lid above ground level and prevents dirt and debris from falling in.

The sump lid should be a well-constructed solid platform that can be easily opened for observation or maintenance and can be locked closed for safety. The sump lid is a valuable working space for harvesting and washing produce so it needs to be able to handle the weight of people moving and working on it.

Water returning from the troughs is plumbed through the east end of the sump. The heat pump, located just outside the south wall of the greenhouse, draws water from the southeast corner of the sump, heats or cools the water as needed and returns it to the sump downstream (farther west in the sump).

If you install a cistern, its intake and return plumbing should be connected near the middle of the south wall of the sump.

The sump is oxygenated constantly via air stones connected to the aerator. The primary purpose of the sump aeration is to vigorously mix all the water in the sump to ensure temperature is consistent and that the pH adjusting chemicals are well mixed before the water is pumped to the fish tanks.

pH adjustments are made in the sump by an automatic pH monitoring and dosing controller (see Chapter 6). The controller is located near the UV sterilizers at the west end of the sump so that its probe can read the pH of the water near the pump intakes. The injection tubing that carries the pH adjusting chemicals from the dosing pump should be suspended over the sump, near the water return line from the hydroponic troughs. This will ensure the chemicals are well mixed and improves the accuracy of the dosing system.

The Drain Down Effect

In DWC systems, the drain down effect occurs when the water pump is shut off following normal system operation. Typically this is because of a power outage or pump failure. The drain down effect is due to the fact that water is "retained" throughout the system by various restrictions such as pipe size, friction and surface tension.

For example, at the east end of each trough is a pipe that carries the water from the north side to the south side. This pipe is a potential bottleneck in the flow. If it is not oversized, during normal operations the water level in the north sides will be slightly higher than the level in the south sides.

A second example is the meniscus (the curve in the upper surface of the water) that occurs at the lip of the trough drain standpipes which is due to the surface tension of the water. Both the trough height differential and the meniscus height at the standpipe are increased at higher flow rates or with smaller pipe sizes.

Although these heights are relatively small, when multiplied by the surface area of a commercial-sized DWC trough, they equate to a fairly large volume. This differential is not an issue during normal operations, but when a power outage or pump failure occurs, all the retained water in the system will drain into the sump. We call this the "drain down effect."

If the sump does not have enough capacity to handle the extra volume, it will overflow and the lost water will need to be replaced before the system becomes operational again.

There are four options to prevent the sump from overflowing:

1. Build a sump that is large enough to hold the drain down volume of the system.
2. Oversize all plumbing connections in the troughs to minimize water retention (inlets, outlets and standpipes).
3. Use a cistern connected to the sump as an overflow collection.
4. Use an automatic standby generator to keep the system operational through a power outage (this will not assist in the event of a pump failure unless your second pump is installed to automatically engage).

Workbench

The workbench will be where you do all the major jobs for plants, notably transplanting and harvesting. The area can be built to your own design, but it should be a comfortable working height (usually waist high) and must have a waterproof top. The easiest method of construction is using standard lumber and plywood that is sealed with a water-based, no VOC, non-toxic sealant. For a trough system, the table must be at least 4′6″ long and 2′6″ wide to accommodate a 2′×4′ raft. For a tower system, the table must be at least 8′ long to accommodate planting into 5′ towers.

Cistern

A cistern is a very valuable and recommended addition to an aquaponic farm for a few reasons:

1. It can capture and store rainwater, lowering your ecological footprint. It is especially helpful if you face seasonal water restrictions or simply want to decrease your reliance on municipal water.
2. It can temporarily store water from your system while performing repairs or maintenance or while performing a treatment protocol for fish health (see Chapter 8).
3. It can be used as overflow for the system to prevent sump overflow during drain down events.

On our farm we use an 8,000-liter cistern.

Power Consumption

Aquaponics is high-tech food production and requires electricity. If possible, a dedicated service of at least 100 amps to the greenhouse is ideal to both guarantee adequate capacity and for ease of accounting. See Chapter 4 for electrical installation.

3

Principles of System Design

IF YOU OPT TO BUILD OUR SYSTEM with a different greenhouse dimension than 36′×120′ or you opt to use a different system or alternate components from the design in this book, you will need to create your own system design. If you choose to design your own system, you should understand it is a complex undertaking and that the concepts and formulas presented here are essential but only scratch the surface of the science behind designing a successful aquaponic ecosystem. Additional research is mandatory for designing your own system.

The Golden Ratio of Cold-water Aquaponics

The design of any size and shape of an aquaponic system starts with a simple ratio which we call the Golden Ratio of aquaponics:

$$G^f/m^2/Day$$

Where G^f is the total grams of feed given daily to the fish and m^2 is the total Hydroponic Surface Area (HSA) in square meters. HSA is the total surface area devoted to plant growth. The purpose of the Golden Ratio is to balance the daily amount of feed that is added to the system with the waste removal capacity of the system and nutrient demands of the growing plants.

The capacity of an aquaponic system to remove waste is determined by two factors:

- The capacity of the biofilter to oxidize ammonia (nitrification), which is determined by the Bacterial Surface Area (BSA).
- The capacity of the hydroponic subsystems to assimilate nitrates (denitrification), which is determined by the Hydroponic Surface Area (HSA).

Nitrification is the process of oxidizing ammonia into nitrates by the nitrifying bacteria. Denitrification is the removal and assimilation of nitrates by the plants. Essentially, with the Golden Ratio, you are balancing the feed input with nitrification and denitrification.

The Golden Ratio was developed at the University of the Virgin Islands specifically for DWC systems which generally have a low BSA/m^2 of HSA (less than 2 m^2 of BSA per m^2 HSA) without taking into account the BSA of plant roots. The plant roots can double the BSA of a DWC system.

Although tower systems can have far greater BSA/m^2 (up to 150 m^2 of BSA per m^2 HSA), both systems have a similar total HSA and thus a similar capacity for denitrification. Accordingly, the Golden Ratio should be used to design both DWC and tower-based aquaponics systems.

Cold-water vs Warm-water Aquaponics

Pioneering studies at the University of the Virgin Islands concluded that the ideal ratio for an aquaponics system is about 60 grams of feed per square meter of HSA per day, or $60g/m^2/Day$.

This appears to work well for warm-water systems operating at an average water temperature of 25°C (77°F), but in a cold water system operating at 15–17°C (59–63°F), the nitrifying bacteria grow much more slowly, thus more surface area (BSA) is required for sufficient oxidation of ammonia. Additionally, nutrient uptake in plants is slower in colder water, and therefore denitrification is also slower. For these reasons, considerably less feed should be given per m^2 of growing area in a cold-water system than in a warm-water system.

IMPORTANT

The Golden Ratio for cold-water aquaponics is $25–35g/m^2/Day$, or 25–35 grams of feed per square meter of HSA per day.

Using the Golden Ratio

The Golden Ratio ensures that an aquaponic system is balanced correctly. There are two primary ways to use this ratio, discussed below as Step 1 and Step 2.

Step 1 starts with a known HSA and uses the Golden Ratio to calculate the daily feed amount and the total fish biomass that can be hosted in the system based on the recommended daily feed rate for the fish you choose to raise. Feed rates for fish are always specified as a percentage of body weight per day. You can use this figure (target fish biomass) in Step 2 to determine the appropriate volume of your fish tank(s) and thus how much water must be delivered to each tank so that the water in each tank is replaced every 30–45 minutes.

Step 1 is used when the initial known variable is the size of the greenhouse and thus the available HSA that will fit in it. This is the most common starting point for designing a system.

Step 2 starts with either a known biomass (target amount of fish you want to grow) or known combined volume of fish tanks (useful if you already have tanks and want to design your system around them). You can then use Step 1 to calculate how much HSA you will need and thus how big your greenhouse must be.

Step 1

The most common application is to start with the biggest greenhouse you can afford, fit on your site and want to operate (for example, you may have a huge property and lots of money but not want to operate a large facility). From this, determine the size of the hydroponic subsystem, and thus the total HSA. In a DWC system, the HSA is the surface area of the rafts. In a tower system, the HSA is best described as the area of floor space dedicated to the tower arrays, not including the walkways.

For example, in a DWC system in a 120′ greenhouse with 86′ troughs (per our design), we produce the following equation:

$$43 \text{ (rafts per side)} \times 6 \text{ (sides)} = 258 \text{ (rafts)} \times 0.743 \text{ m}^2 \text{ (size of each raft)} =$$
$$191.69 \text{ m}^2 \text{ (HSA)}$$

For our calculations, the HSA for our design is 192 m^2.

We then use the formula to find the daily feed rate, following the Golden Ratio:

$$25 \text{ grams} \times 192 \text{ m}^2 = 4{,}800 \text{ g or}$$
$$4.8 \text{ kg of feed per day}$$

$$30 \text{ grams} \times 192 \text{ m}^2 = 5{,}760 \text{ g or}$$
$$5.76 \text{ kg of feed per day}$$

$$35 \text{ grams} \times 192 \text{ m}^2 = 6{,}720 \text{ g or}$$
$$6.72 \text{ kg of feed per day}$$

This shows that the amount of feed that can be safely added to the system each day without overwhelming the nitrifying bacteria or allowing nitrate levels to rise excessively is 4.8 kg to 6.72 kg. With this knowledge, we can figure out the total fish biomass that can be grown on this amount of feed based on the daily feed rate for a specified species of fish.

Rainbow trout, which we raise in our system, consume an average daily feed rate of 1.3% of body weight (see Chapter 8), therefore:

@ 25 g/m²/day:
4.8 kg / 0.013 = 369 kg total biomass

@ 30 g/m²/day:
5.76 kg / 0.013 = 443 kg total biomass

@ 35 g/m²/day:
6.72 kg / 0.013 = 517 kg total biomass

Thus after calculating the range of feeding ratios (from 25–35 g per m² per day), we know that we have a target biomass range of 369 kg to 517 kg.

The next step is to calculate how to divide the total biomass into multiple cohorts. This would be simple if just using one tank and growing one cohort at a time. Unfortunately, raising single cohorts is not acceptable in an aquaponic system as it is critical the system be consistently loaded within its range: too little fish biomass will not adequately feed the bacteria and plants, too much biomass will overwhelm the bacteria and harm or kill the fish. A single cohort system would be severely underloaded when the fish are fingerlings and could be overloaded when the fish are mature, depending on how large your fish grow.

To ensure a consistently balanced system, fish are grown in multiple cohorts of different sizes in separate fish tanks, and when the system is fully loaded, small batches are harvested from the oldest cohort every 1–3 weeks.

We can now use Step 2 to determine how many fish should be in each cohort while staying within the Golden Ratio.

Step 2

Use Step 2 if you already know the biomass range for your system, or the size of tanks you want to use, or to ensure that your design has sufficient HSA to support your biomass range. Continuing our calculations from Step 1, the process is as follows.

Our design uses three 8′ tanks, each with a volume of 4.2 m³. The recommended maximum stocking density is 60 kg/m³, therefore each tank can carry a maximum of 250 kg of fish at their target harvest weight:

60 kg × 4.2 m³ = 250 kg max biomass (rounded down from 252 kg)

Fish can be harvested at different weights for different markets. If the target harvest weight for each fish is 1 kg, you should stock each 8′ tank with 250 fish. If the target harvest weight is .5 kg, stock 500 fish per tank. Consider your potential markets when deciding on target harvest weight.

Let's assume that 1 kg is the target harvest weight and that you will be restocking every four months with 250 fingerlings at 10–12 cm long and about 10–20 g each, which is what we do at Raincoast. At 15°C (59°F) water temperature with ideal water and feed conditions, rainbow trout are capable of growing 3 cm per month. If the fish start at an average of 12 cm long when they enter the system, then after 4 months they should average 24 cm long:

$$12 \text{ cm} + (3 \text{ cm} \times 4 \text{ months}) = 24 \text{ cm}$$

The average weight of fish is mathematically related to their length, using a formula called the Condition Factor (CF). The CF of rainbow trout under normal conditions is 1.11. You can predict the average weight of the fish at any given time using this formula:

$$\text{Weight(g)} = \text{CF} \times (\text{length[cm]})^3 / 100$$

In our example, the estimated average weight per fish after 4 months of growth is:

$$\text{Weight} = 1.11 \times (24)^3 / 100$$
$$\text{Weight} = 1.11 \times 13,824 / 100$$
$$\text{Weight} = 153 \text{ grams per fish}$$

After 8 months:

$$\text{Weight} = 1.11 \times (36)^3 / 100$$
$$\text{Weight} = 1.11 \times 46,656 / 100$$
$$\text{Weight} = 517 \text{ grams per fish}$$

After 12 months:

$$\text{Weight} = 1.11 \times (48)^3 / 100$$
$$\text{Weight} = 1.11 \times 110,592 / 100$$
$$\text{Weight} = 1.2 \text{ kg per fish}$$

With these calculations, we can determine the total biomass per tank at different stages:

Tank 1 (4 months):
250 fish × 153 g each = 38 kg

Tank 2 (8 months):
250 fish × 517 g each = 129 kg

Tank 3 (12 months):
250 fish × 1.2 kg each = 300 kg

Total fish biomass = 467 kg

Now we have a snapshot of the total fish biomass estimated to be in the system when operating at full capacity. When the system is running at full capacity, you should be harvesting fish on a regular basis from the oldest cohort to keep the total biomass of the system within the acceptable range. Accordingly, the total biomass will often be less than the maximum calculated here. A Cohort Log for each cohort of fish will help you keep track of the biomass in each tank at any given time, and these logs will tell you how much feed should be given to each tank daily. Fish Sample Logs and Cohort Logs are covered in Chapter 11.

A total biomass of 467 kg is within the target range of 369–517 kg and within the range of the recommended Golden Ratio:

467 kg × 1.3% (daily feed rate) = 6.07 kg feed per day

6.07 kg / 192 m^2 HSA = 31.6g/m^2/day

By using the formulas given above, we determined that with 192 m^2 of HSA the system can carry 467 kg of fish and that the required daily feed input of 6.07 kg is well within the acceptable range of the Golden Ratio of 25–35g/m^2/day. We also know that each 8′ tank can carry a cohort of 250 fish without exceeding the maximum recommended density of 60 kg/m^3.

As a further exercise, we can use the Golden Ratio and the daily feed amount to calculate the minimum and maximum HSA and thus the number of rafts as follows:

Minimum:
6.07kg / 35g = 173 m^2 HSA = 232 rafts = 78′ troughs

Maximum:
6.07kg / 25g = 242 m^2 HSA = 325 rafts = 108′ troughs

A Note on Tower Systems

These design principles apply to both DWC and tower systems: the feed input and ammonia generated needs to be balanced with the system's capacity to remove nitrates before they build up to levels toxic to the fish. The major advantage of ZipGrow™ Towers, which use a matrix media, is that they have a much greater BSA per square meter of HSA. They are therefore more forgiving of mistakes like overfeeding or infrequent cleaning of filters because they convert the ammonia to nitrates much more quickly than DWC systems. Other types of tower systems, such as those filled with gravel, have a much lower BSA than ZipGrow™ Towers.

A Final Design Note

If you opt to design and build an aquaculture system that is a major departure from the design presented in this book, we wish you the best of luck. Designing a fish culture system that will keep your fish not only alive but thriving will be the most challenging and scientifically complex part of the design. Before you embark on that journey, we highly recommend that you read and understand the book *Recirculating Aquaculture* by Timmons & Ebeling (3rd Edition, Ithaca Publishing Company, 2013). It is the definitive text on recirculating aquaculture systems and likely contains everything you need to know about keeping your fish alive. It is a highly technical book, and we have just scratched the surface here. It even has a chapter specifically on aquaponics. Whether you design your own system or follow our design, we recommend it as a reference book for all aquaponic farmers.

Constructing the RCA System

4

In Chapter 2 we covered all the major components and elements of our design. Additional components are in Chapter 5. This chapter contains the literal nuts and bolts of constructing and installing the system, from site preparation to being ready for water to flow through the system.

IMPORTANT

Unless you are a professional contractor with considerable construction skills and experience working with various trade contractors, we strongly suggest you hire a general contractor (GC) to oversee your construction. A good rule of thumb is this: if you would hire a GC to build a house, you should hire one to build your greenhouse and aquaponic system. Many of the instructions we provide in this chapter are general in nature, and the construction and/or installation may require a professional builder. Every site will have its own unique demands. We cannot cover the myriad contingencies and problems you will encounter that are specific to your site and project. For all tasks, follow your local building code as applicable.

Always follow manufacturers' installation and operation instructions. If our instructions or recommendations contradict a manufacturer's, follow the manufacturer.

The general order for construction is:

- Site Preparation
- Greenhouse Construction
- Electrical Installation
- Sump Construction
- Waste Tank Excavation
- Trough Construction
- Trough Liner Installation
- Trough Plumbing Installation
- Aquaculture Subsystem Installation
- Pump and Plumbing Installation
- Heat Pump Installation
- Cistern Installation
- Seedling Table Construction
- Workbench Construction
- Aeration Installation
- Cooler Installation
- Germination Chamber Construction

Site Preparation

We have previously covered the various considerations for choosing a property and buying a greenhouse. By this point, we assume you have secured a property, ordered a greenhouse and obtained all necessary permits to build. In securing your property, you will have chosen your specific greenhouse site based on the factors covered in Chapter 2.

The first order of business is to excavate the site. Have an excavator level the ground by removing all turf, topsoil and compostable matter, and scrape down to undisturbed, compacted earth. The area required is a minimum of 10′ larger than your greenhouse on all four sides (140′×56′ for a 120′×36′ greenhouse). After the initial excavation, use a laser level to confirm that the area is level to within a few inches. All large rocks, stumps or debris must be fully removed if exposed above this level.

Greenhouse Construction

If you are buying a new greenhouse, it will come with detailed installation instructions as well as all required hardware. Follow the instructions to the letter.

If you are buying a used greenhouse, one of the biggest downsides is it is likely that it will not come with installation instructions. Accordingly, you must either be meticulous in documenting how the greenhouse is taken down so you can reverse steps to reinstall it, or you should hire professional installers.

Foundation Installation

Most greenhouses will require either a pier foundation or a stem wall foundation. A pier foundation costs less (less concrete) and is easier to install as the piers do not require forms or framing. The foundation posts are set into the freshly poured concrete piers to support the arches. The arches will bolt onto the foundation posts.

Using batter boards and string lines, lay out the piers per the greenhouse specs. Mark the string line with the position of each foundation post that will be set into the concrete pier. Using an auger, dig the pier holes to the diameter and depth specified by the manufacturer. Fill the holes with concrete and place the foundation posts as the concrete sets, using string lines as a guide for accuracy.

Dig a perimeter drain around the greenhouse. If there is a local building code that applies, follow the code. The perimeter drain must be at least 2′ deep and 2′ away from the piers. Use Schedule 40 perforated 4″ PVC that drains away from the greenhouse. Backfill the perimeter drain with 12″ of

Foundation installation with string lines and batter boards, ready for concrete piers.

drain rock, cover the rock with landscape fabric to prevent dirt clogging the pipe, then fill to ground level with gravel.

Arches Installation

The concrete piers should be at least seven days old prior to installation of the arches, so that the concrete is mostly cured. We recommend having at least four strong people onsite for this job. Follow the manufacturer's instructions carefully.

In general, the procedure is to construct each arch while it is lying on the ground and then stand it up using muscle and rope. If possible, rent a scissor-lift or a telescoping forklift to assist with standing up the arches.

The arches are shipped in halves and bolted together onsite with cross braces. Start at one end of the greenhouse and work your way to the other end, standing up one arch at a time. After you install the second arch to its piers, bolt it to the first with the supplied purlins. After you install the third arch to its piers, bolt it to the second, and so forth. Once all arches are in place, add the wind bracing, which will greatly increase structural integrity. When complete, the arches will resemble the ribs of a large metallic creature or an upside-down boat.

Before building the endwalls, confirm that your fish tanks will fit through the large sliding doors. If they will not fit, move the tanks into the greenhouse.

Greenhouse frame.

Endwall Installation

The endwalls of greenhouses vary considerably. Follow the instructions carefully. Part of the installation will be installing the power vents, fans and doors that are housed within the endwalls. Plan out where the electrical, data, water and propane services will enter below the endwalls and install conduits for them.

Typically, polycarbonate endwalls (as we recommend) are installed by first constructing a steel framework of vertical posts and horizontal braces then securing the sheets of polycarbonate to the frame. The wall framing and bracing will vary from greenhouse to greenhouse depending on the layout of doors, vents and other components.

After the polycarbonate is secured to the wall, metal rain flashing is fit over the joint between the roof and endwall, and then C-channel or another system for securing the roof covering is installed over the flashing.

Once the endwalls are installed, install baseboards around the perimeter of the greenhouse.

Covering Installation

Next, install the greenhouse covering. If you are buying a new greenhouse, you should receive detailed instructions on how to install your chosen covering. Follow the instructions carefully.

Polycarbonate endwall.

If you are using double-layer poly, install the channels or other mechanism that will secure the sheets to the greenhouse, typically metal C-channel running around the perimeter. The poly is tucked into the channel and a wire or clip is used to lock it in place.

Poly will come on a large hollow-core roll and can be installed by various methods. The easiest is to roll out both layers on another sheet of clean plastic on the ground along one side of the greenhouse. Throw three ropes over top of the arches, one near each endwall and one in the middle of the greenhouse. Attach the ropes to the corners and middle of the poly, making sure each rope is secured to both sheets at the same time. With one person on each rope, slowly drag the sheets of poly over the greenhouse, taking care not to tear them on sharp objects. Once the sheets are straightened out and free of wrinkles, they can be secured to the C-channel and any excess trimmed.

Roll-up Sides Installation

Roll-up sides consist of a tube running the length of the greenhouse, around which the bottom edge of double-layer poly is wrapped and secured. The top of the poly is secured to C-channel that runs along the side purlin, typically about six feet above ground. A geared cranking handle at one end of the tube is used to roll-up the tube and both layers of poly with it. In the winter, both ends of the roll-up sides are secured to C-channel on the two end arches, and the space between the layers of poly is inflated using the same blower that keeps the covering inflated.

Hanging Components Installation

Mount all items that will be hung from the cross braces of the greenhouse arches. This includes the propane heater, circulation fans and hanging supports for the HID lights (the lights will be installed later). Follow the installation instructions for the heater and the circulation fans.

Circulation Fans Installation

The circulation fans should come with a specific layout and direction of flow that maximizes circulation and mixture of air. In general, the air will flow in a circle around the greenhouse. Our 80′ greenhouse has four circulation fans.

Heater Installation

Prior to installing the heater, we suggest consulting with the gas fitter you will be using to run your propane line. Discuss placement of the heater, venting options, where the propane tank will sit and the path the gas line will take between tank and heater.

If you are using a double-poly covering, you must vent through the end-wall closest to the heater. Our heater is located in the northwest corner above the Seedling Table. As discussed in Chapter 2, it is advantageous to use rigid polycarbonate for the endwalls as it allows lines and ducts to pass through neatly and easily.

If using a propane or gas heater, you will require a large storage tank onsite. These can either be purchased or rented on an annual basis from the company that will supply your propane. Ours costs about $10 per month to rent. The service company comes to our site and fills it on request.

HID Light Installation

Supplementary light will be required as discussed in Chapter 2. We use 1000W HPS (High Pressure Sodium) lights that have lamp, ballast and reflector in one integrated unit. Our lights are on light movers, with each bulb responsible for covering approximately a 20′ long span of each 8′ wide trough. For a 120′ greenhouse with 86′ troughs, you will need four lights per trough, twelve lights total.

There are numerous DIY methods for hanging your lights. For a trough system, the lights will move directly over the middle wall in each trough. For a tower system, install lights as specified by the tower manufacturer.

Our simple system for hanging lights is:

1. Hang chains from the cross braces over the center of each of the three troughs.
2. Install S-hooks at the bottom of each chain.
3. Hang wood 2×3s (or 2×4s) on edge for the length of each trough, using eye screws in the top edge for the S-hooks. The 2×3s can be replaced with metal if you prefer. Make sure the eye screws are rated to hold 100 lbs.
4. Prior to hanging, screw the light mover tracks into the bottom edge of the 2×3s. Make sure to accurately align the ends of the tracks to ensure smooth movement of the light movers.

At this point in construction, the only mandatory step is to hang the chains, as this is more difficult once the troughs are installed. Steps 2–4 can be done either now or later when you install the lights.

We use and highly recommend Gavita Pro 1000W DE units for the troughs. We also use and recommend LightRail light movers.

Electrical and Internet Installation

Unless you are an electrician by trade, you should hire an electrician. Improperly installing any other component in the system can lead to crop failures, dead cohorts and lost income, but improperly installed electrical can kill you. Follow code as applicable and listen to the advice of your electrician.

For ease of installation, the electrical cables should be run from the main electrical panel to their outlets before installing the troughs or fish tanks. The main service connection to your greenhouse will run either overhead from a pole near the greenhouse or underground to the main electrical panel. However the power is brought into the greenhouse, we strongly suggest running cables in the greenhouse overhead in the arches instead of burying them. Use PVC electrical conduit or as specified in your electrical code. All conduits, connection points and junction boxes should be moisture-proof PVC, and all outlets must be GFCI.

The location of the main panel will depend on local codes and where the main power for the property is located. If you opt to install the panel inside the greenhouse and code allows this, the box should be installed on an endwall. Most of the circuits will be at the west end of the greenhouse, but this endwall will also become very crowded. Note that if you install the panel on the east endwall, you will likely need to oversize some of the cables to account for voltage drop due to the length of the circuits.

Our power enters via a buried cable from the east. Our panel is located on the inside of the east endwall, and the cables run overhead.

The following are suggested circuits listing components that can be grouped together or components that require dedicated circuits. In some cases, these suggestions may not be possible if the voltage for the component is different than shown here (e.g., if the water pumps are 240v and the UV units are 120v).

- Water pumps and UV sterilizers
- Aeration blower
- Trough HID lighting (this may require multiple circuits)
- Seedling HID lighting
- Heat pump
- Walk-in cooler condensing unit
- Walk-in cooler evaporator and Germination Chamber
- Auxiliary outdoor power (washing machine, effluent pump, etc.)
- Greenhouse circulation fans and propane heater
- Endwall ventilation fans and poly inflation blower
- Plug outlets by stainless sink and emergency oxygen system
- Plug outlets by sump for water monitoring equipment (pH controller, Web600)

- Plug outlets by workbench area
- Two plug outlets on the east endwall (one at the end of each middle walkway)

Internet Installation

The monitoring system (see Chapter 5) will likely require a stable internet connection to send alarms. We highly recommend running Cat5e cable rather than trusting any Wi-Fi system. In most cases, the ideal time to run the Cat5e cable is in a conduit in the same trench as the buried power cable.

Sump Construction

IMPORTANT

Safety around the sump is a major concern. Your sump will be a 4′ deep hole filled with water. If a child or pet were to fall in, they could easily become trapped and drown. Be mindful of this and take precautions during construction. It is mandatory that you have a locking lid on your sump once constructed.

Our design calls for a sump that is 4′ wide, 10′ long and 4′ deep. These dimensions are all exterior, meaning once you have built the sump walls, the interior dimensions will be about 3.5′ wide, 9.5′ long and 4′ deep, with a total volume of 170 cubic feet or 4,800 liters.

If your site does not allow the sump to be a depth of 4′, compensate by extending the length but not the width. Expanding the width towards the greenhouse wall may compromise the adjoining pier foundations. Expanding the width towards the fish tanks will compromise your walking and working space. Consider also the dimensions of the material being used to line the sump. If using LDPE as a liner, the depth and width of the sump are limited to the size of the liner.

Sump Construction

Using non-toxic spray paint, mark a spot 17′ from the west endwall and 2′ from the south sidewall. This is the southwest corner of the sump. From this corner, mark out a rectangle 4′ to the north and 10′ to the east. Double-check the layout is square by measuring both diagonals.

Excavate the hole at least 4′ by 10′ by 4′ deep at all points. Fill the hole with a 2″ layer of bedding sand to prevent sharp rocks from piercing the liner and tamp it flat. You will also use a 2″ layer of bedding sand as the base for your troughs, fish tanks and cistern. You can save delivery fees by calculating your total sand requirement and ordering only once.

The walls will need to be shored up to prevent collapse over time. There are several methods to do this, including treated timbers and plywood, poured concrete walls or burying a rigid poly tank. It is not acceptable to simply line a bare dirt hole. Using concrete or rigid poly has the advantage of longevity but is very expensive and in-wall service connections are very difficult.

Our recommended method is to construct 4′ high stick frames walls, similar to those used in houses, using treated 2×3s sheathed on both sides with ½″ treated plywood. The studs should be on 12″ centers. Use only exterior screws as fasteners.

Wood in a constantly moist environment (such as in contact with the ground) will eventually rot. Expect buried treated lumber and ply to last 5–10 years before needing repair or replacement. Expect non-treated lumber and ply to last 2–5 years before needing repair or replacment. Also note that using treated lumber may compromise organic certification.

Sump. Side view.

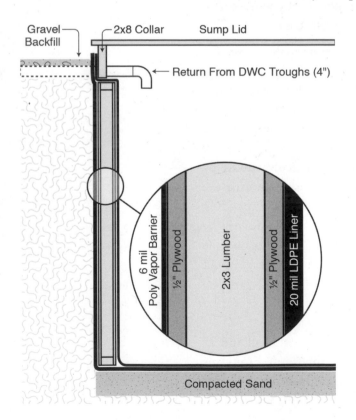

Prior to installing the walls, line the sump with a 6 mil poly vapor barrier to prevent wood deterioration. Do this even if you are using treated lumber and plywood.

Install the walls in the sump and fasten them together at each corner with exterior screws.

Line the sump with 20 mil LDPE. This is the same material you will use to line the troughs, so make sure to order enough for both. If you choose to use LDPE to line the Seedling Table or workbench, include these amounts in your order as well. You must use a liner regardless of the type of walls you've built in your sump, including concrete. Concrete is not waterproof and will leach horrible things into your system water.

Wrap the liner over the top of the walls before installing the collar so that the liner is pinched between them.

On top of the walls, fasten a collar of 2×8 lumber on edge. The collar will support the lid and raise it above the ground, preventing debris from entering the sump. The top of the collar should be 3″ above the final ground level (including gravel), so plan the height of the walls accordingly. Prior to installing the collar, apply two coats of epoxy resin to all six sides of each board and allow them to cure.

The sump will be a constantly wet area. All exposed wood, including the collar, lid support joists and lid, must be sealed with two or three coats of epoxy resin. Paint alone is not sufficient.

Once they have been coated and cured, install horizontal joists across the width of the sump that will support the lid. Install a joist every 2′. The joist nearest the west end of the sump should be a double joist, as one side will support the sump lid that opens and the other will support a fixed plywood cover around the pump intakes.

Install the sump lid with non-rusting hinges of your choosing and the fixed plywood cover with exterior screws. The simplest and recommended method for the lid is to use a single sheet of ¾″ untreated plywood, water-proofed with epoxy resin. Apply two coats of resin to the sump lid and fixed plywood covering. We recommend three coats for the top of the lid as it will be walked on. Add a small amount of sand while applying the resin to the last two coats on the lid for a non-slip top. Optionally for aesthetics, you can paint the top of the sump lid after applying resin with a water-based, no VOC, non-toxic, non-anti-microbial paint (the same paint used to paint the rafts). Install a locking latch on the sump lid for safety.

In a tower system with two split loops, additional sump design consideration is required to ensure effective mixing of the water from both loops. Install the fish tank pumps at the southwest corner and the 4″ return line

Sump. Top view with cover removed.

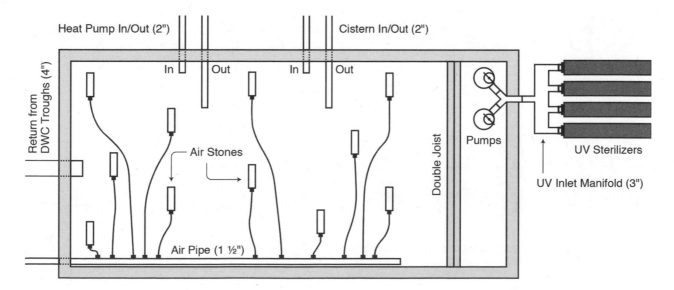

Heat Pump In/Out (2″)   Cistern In/Out (2″)

Return from DWC Troughs (4″)

In   Out   In   Out

Air Stones

Double Joist

Pumps

UV Sterilizers

UV Inlet Manifold (3″)

Air Pipe (1 ½″)

at the northeast corner (same as DWC design). Install tower pumps at the southeast corner and the return line at the northwest corner.

If possible, all screws used should be stainless steel. Even exterior (coated) deck screws will rust eventually if they are constantly exposed to moisture. This applies for all construction but is particularly important in the sump.

IMPORTANT

The two parts of an epoxy are the resin, which is non-toxic, and the hardener, which is very toxic. If the two parts are not mixed in proper proportion, there is a possibility of uncured hardener in the mix which can leach and be toxic to fish. Make a resin-rich mix (more resin than hardener) to avoid this problem. Never use polyester resin. **This is important for all epoxy applications.**

Waste Tank Excavation

We suggest using one or more standard 1,000-liter IBC totes for waste collection that will be located at least 20′ outside the greenhouse to the west. If there is no elevation change at your site, you will need to dig a waste collection

Plumbing for waste collection. Note the requirement for elevation drop to Waste Collection Tanks.

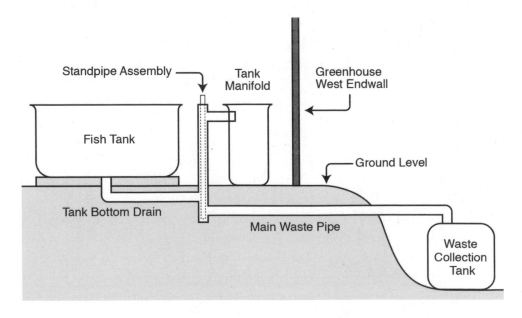

pit. The pit should be big enough to house the Waste Tanks with their tops below the outlet level of the bottom of your fish tanks, including a 1″ drop every 10′ of pipe length.

We suggest digging the pit while you have an excavator onsite. However, do not install the Waste Tanks yet. This will be done after the underground plumbing is roughed in. If in doubt about how deep to dig the pit, dig it 2–3′ deeper than you think is needed. Pack a base of gravel for the tanks to sit on and, if possible, create a drainage trench for when they inevitably overflow.

Trough Construction

Ground Preparation

The first task is to prepare the ground in the 90′-long area where the troughs will be located (86′ for troughs plus a 4′ walkway at the eastern end of the greenhouse). The ground must be flattened and compacted to within ¾″ over the width and length of the area. It is very important that the area be level to this accuracy or the troughs will end up warped. Spread 2–3″ of ¾″ crushed gravel over the area and compact it with a heavy plate compactor until it is as flat and hard as possible. Use a surveyor's level for accuracy.

Lay out the rough location of the troughs on the ground by marking their corners with a water-based spray paint. Start by lining up the middle of the center trough with the center of the greenhouse by hanging plumb bobs from the center of the cross braces on two arches, one at each end of the trough. Mark the location of the middle wall of the center trough, then work outward from the center, measuring and marking the aisles and the corners on both ends of the troughs. Finally, mark the middle wall locations of the two outer troughs.

The outside dimensions of the troughs are 86′8″ long and 8′9″ wide. In a 36′ wide greenhouse, the walkways are 29″ wide. In a 40′ wide greenhouse, the walkways are 41″ wide. The eastern edges are 44″ from the east endwall of the greenhouse.

At this time your objective is to mark the layout close to the final positions, but you don't need to be precise. Once the walls of the troughs are constructed and in their approximate location, string lines are used to accurately place them in their final location and square them.

Trough Construction

Each of the five walls of a trough is 13″ high. The length of the two side walls and one middle wall are each 86′4″. The two shorter endwalls are 8′9″ each. All walls are constructed of standard ½″ plywood and 2×3s, except the bottom plates which must be treated 2×3s or standard 2×3s stained with a wood preservative paint. We recommend pressure treated for the bottom plates.

Trough construction consists of the following steps:

1. Constructing 6 side walls, 3 middle walls and 6 endwalls
2. Assembling the walls
3. Final placement of the troughs
4. Trough liner installation
5. Plumbing installation

Side and Middle Wall Construction

Construct all walls in the approximate location they will be positioned so that they can be simply stood up and screwed together. Do not attempt to move constructed walls long distances as they are very heavy and awkward.

To build the side and middle walls:

- Lay out nine 10′ 2×3s, end to end. Cut the first 2×3 to 76″. This will be the top of the wall.
- Lay out a second row of nine 2×3s parallel to the first. Cut the ninth 2×3 to 76″. This will be the bottom of the wall.
- Cutting the opposite ends to 76″ will stagger the butt joints. This is very important.
- Cut 8″ pieces of 2×3 blocking and place at both ends and at every butt joint on both the top and bottom plates.
- All 2×3s in the frame will be on edge.
- Fasten all 2×3s together using exterior screws to assemble the frame which should be 13″ tall by 86′4″ long when complete.
- Cut sheets of ½″ plywood lengthwise creating four 12″ by 86″ pieces from each sheet. Don't worry about the slight loss of width due to the saw kerf.

Trough side wall construction.

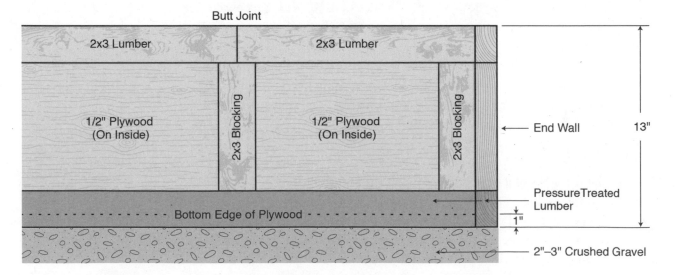

IMPORTANT

The plywood is on the inside of each side wall. It is critical to leave the 1″ gap at the bottom of the frame so that, when stood up, it will be towards the middle of the trough. If you do the opposite, you will have to spin the wall lengthwise which may prove impossible.

- Sheath the 2×3 frames with the pieces of plywood on the inside of the frame, aligning the plywood with the top edge of the 2×3 frame so that there is a 1″ unsheathed strip along the bottom edge of the frame. Use exterior screws and avoid having joints in the plywood line up with the joints in the 2×3.
- Repeat the framing and sheathing process eight more times for the rest of the long walls.
- The middle walls are constructed exactly the same way as the sides, except it does not matter which side the plywood is on. You do not need to sheath both sides of the middle wall.

The 1″ unsheathed gap on the bottom plates serves two purposes: plywood doesn't touch the ground so it does not need to be treated, and the gap will serve as a depth and level guide when you fill the bottom of the trough with sand.

Endwall Construction

The endwalls are 105″ long. The frames are built in the same manner as the side and middle walls: 13″ high frames constructed from 2×3s with bottom plates pressure treated and 12″ high pieces of ½″ plywood. There must be a vertical 2×3 centered in the middle of each endwall in order to fasten it to the middle wall. Do not sheath the endwalls at this time; this is done after the walls have been fastened together.

Assembling the Walls

For each of the three troughs, you should now have two long side walls with ply on the inner sides, one long middle wall with ply on one side and two endwalls that are framed but not yet sheathed, all with 1″ unsheathed at the bottom.

Confirm the rough layout for each trough per the previously made markings so that only minor adjustments are required. The side walls of the troughs will be quite flexible when you initially stand them up. We suggest having five or six people on hand to help with this job.

The side and middle walls butt into the endwalls. Stand one side wall and line it up with the corresponding mark. Stand one endwall at a time and join them to the ends of the side wall. Use at least three ¼"×4" lag screws per butt joint. First drill pilot holes, then countersink the heads by ¼". The trough should now look like a very tall "C".

Stand the middle wall and fasten it to the endwalls using 4" lag screws. The middle wall must be centered on both endwalls. Stand the other side wall and fasten to the endwalls using 4" lag screws.

It is likely that the frame will be floppy until it is shimmed and lined with the poly. If necessary, use temporary 2×2 bracing across the width of the trough to keep the side walls parallel during the final adjustments.

Sheath the inside of the endwalls with ½" plywood, leaving a 1" gap at the bottom.

The frame of one trough is now assembled. Repeat the assembly process for the other two troughs.

Final Placement of the Troughs

Use spikes or short pieces of rebar and string lines to lay out the exact corners of the troughs.

44" in from the east greenhouse endwall, use stakes to mount a string line north-south at 14" high. This is the outside edge of the east end of the troughs. Use a permanent marker to label the string at the center of the middle trough using the plumb bobs previously hung from the cross braces, then label the rest of the trough corner locations working your way outward in both directions.

Measure 86'8" to the west in 3 locations: the southwest corner of the south trough, the middle of the center trough and the northwest corner of the north trough. Stretch another string line north-south at 14" high to mark out the outside edges of the west end of the troughs. Mark the string line at the center of the middle trough and then the corner locations of the troughs, as before.

Use a sledgehammer and a striking block (so you are not directly hitting the frames) to tap the troughs into place, lining them up accurately with the labels on the string lines.

Once all corners of the troughs are in their final positions, use nine string lines stretched east-west from corner to corner as a guide and straighten the side and middle walls by tapping them into place with the sledgehammer and striking block.

It is likely that the troughs will need to be shimmed with wedges or gravel as the compacted gravel base is unlikely to be perfectly level. Take your time doing this and use a surveyor's level to ensure that the tops of the walls are a uniform height.

Trough Liner Installation

We recommend using 20 mil LDPE as trough liner. This material is often called "geomembrane" and is used as a protective liner to prevent runoff in landfills and waste treatment facilities. It may be possible to get a discounted price for offcut LDPE that is trimmed from a larger project. You can use up to 40 mil LDPE, which is what we have in our greenhouse, but it is much more difficult to work with due to its thickness, and in colder temperatures will likely need to be warmed before installation.

Laying the sand base in newly constructed troughs to protect the LDPE.

The liner must be 14′ wide and 4′ longer than your trough length (90′ long for each 86′ trough) to allow a little wiggle room if you don't install the liner perfectly straight. The liner must be white on its upper side (the other side can be any color including white).

Remember when ordering to add enough for your sump, and more if using it for the Seedling Table or workbench.

Before lining the troughs, prepare a 1″ compacted sand base inside them to prevent sharp rocks from puncturing the liner when filled with water. Use a fine bedding or masonry sand that is screened and does not contain any rocks.

Spread the sand evenly inside the troughs and screed it level, using the plywood gap on the trough walls as a guide. Compact the sand using a small plate compactor or a hand tamper. Keep leveling and compacting until as compact as possible, level and flush with the bottom of the plywood.

You should receive the liner on a hollow-core roll. Unroll it over the length of a trough starting at the west end. The best method is to hang the roll from a tractor bucket or forklift and drag the liner over the trough. Leave 2′ overhanging both endwalls. If you purchased one roll with enough liner for all three troughs, be sure to cut the length precisely so you don't leave yourself short.

Mark the center of both ends of the liner and mark the center of the trough on the outside of the endwalls to assist in keeping the liner square on the frame. Push the liner down into the trough along the east endwall, keeping the center marks lined up and leaving 6–12″ of overhang. Bend the liner over the top edge of the east endwall and tack with a few staples to hold it in place.

IMPORTANT

Only tack the liner to the top edge of the trough walls, never below the waterline.

IMPORTANT

Whenever you walk on the liner at any point, you must do so in clean socks only. Never walk on the liner with shoes or you risk puncturing it.

Working from the east end, push the liner down along the inside of each side wall and both sides of the middle walls. Do this by making numerous passes, always starting from the east (tacked) end and working towards the west (untacked) end.

Rolling out the LDPE for liner installation.

Nearly completed trough liner installation.

IMPORTANT

The bottom of the liner along each of the side walls and both sides of the middle wall, where it meets the ground, must have a small curve with a radius of approximately 2–3″. Do not push the liner square into the crease. This curve is needed for the structural stability of the trough once the weight of the water is exerting force. The curve will distribute the weight of the water evenly between the walls and the ground so that the walls are neither pushed out nor pulled in. It is important that the curves be consistent along each wall and in relation to the opposing wall.

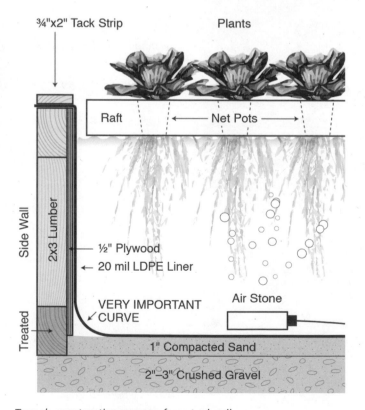

Trough construction as seen from endwall.

Fold the liner at each inside corner as if you are wrapping a present. Some experimentation will be required to learn the best technique. It is necessary to cut slits in the liner to fold the corners but make sure that no cut is lower than the top of the trough once the liner is fully installed. Be conservative with any cuts you make.

The bottom of the liner along each of the endwalls must be pushed square into the crease where the wall meets the ground (no small curve like the side and middle walls). This is necessary for the installation of the bulkhead fittings.

Once you are satisfied with the position of the liner, including the corners and center markings of both endwalls, tack the liner to the top of the frame using a staple gun. Start at the east end where you initially tacked and work your way along the side walls to the west end. Do not tack the middle wall. Remember: you must not wear shoes when walking on the LDPE.

Once you have tacked both side walls and the west endwall, the liner will be in its permanent position. When filling the troughs with water for the first time, fill both sides of each trough at the same time to prevent the liner from shifting. Unless the curves in the liner at the bottom of the side walls are perfectly straight and parallel, which is difficult to achieve, the walls of the troughs may bow in or out slightly once they are full of water. This should not be of concern, unless the bow is significant enough to pinch the rafts or to extend too far into the aisles. If this happens, the bow can be corrected by cementing a vertical support into the ground like a fence post, and securing the side wall to it.

To permanently secure the liner and create a finished edge, fasten ¾"×2" wood tack strips (ideally red cedar) along the tops of the side and endwalls using 1½" exterior screws every 12". The screws will go through the tack strips, the LDPE and into the wall frames. Do not use tack strips on the middle wall.

Using a new blade on a utility knife, trim the excess liner on all four sides.

Paint all exterior surfaces of the trough walls using a white 100% acrylic latex exterior paint. It is mandatory that the paint not contain any mold or mildew inhibitors (such as MicroBan). This is the same paint you will be using to paint the rafts.

Trough Plumbing Installation

Water will flow in the north side of each trough from the west end, take a U-turn at the east end and flow back in the south side towards the sump.

The 3" bulkhead fittings used for the U-turns and drains can be quite expensive and may be difficult to find, but we strongly recommend them. They prevent water retention and minimize the drain down effect that occurs when smaller plumbing is used (see Drain Down Effect in Chapter 2). It is possible to use 2" plumbing for all the trough connections, but we do not recommend this. If you use smaller plumbing, you must install a cistern that is connected to the sump to handle the drain down effect. We recommend installing a cistern no matter what size plumbing is used, but it is mandatory if the U-turn and drain fittings are smaller than 3". The trough inlet fittings can be 2" as this won't increase water retention in the troughs.

IMPORTANT

When using a hole saw to cut through the trough walls, always drill from the inside out as this will result in a cleaner, more accurate hole in the liner.

Inlet Plumbing

At the west end, install a 2″ bulkhead in the center of the north side of the trough, as high as possible (just under the 2×3 top plate). From outside the trough, mark the location of the bulkhead and drill a pilot hole through the plywood and trough liner. Use a hole saw of the correct size for the bulkhead fitting (usually a 3″ hole saw for a 2″ bulkhead but measure before drilling).

Place the body of the bulkhead through the hole with the flange head on the outside of the trough. Bulkhead fittings in the troughs are always installed with the flange head on the outside to save space in the walkways. Slide the gasket over the threaded body inside the trough and thread the clamp ring on until it compresses the gasket against the trough liner.

Thread a 2″ MNPT x slip adapter into the flange head and install a 2″ ball valve as close as possible to the bulkhead on the outside. The valve will allow you to adjust the flow to the trough or shut off the water completely.

Drain Plumbing

The trough drain standpipes control the height of the water in the troughs.

Install a 3″ bulkhead for the standpipe drain on the west end of the south side of the trough as low as possible (just above the 2×3 bottom plate) with the flange head on the outside. Typically this will require a 4″ hole saw; measure before drilling. Thread 3″ MNPT x slip adapters into both sides of the bulkhead.

Using two short pieces of 3″ PVC and one 90° elbow, make a standpipe drain inside the trough. The pipe should extend to the top of the trough walls. You will cut it down to final height later. Do not glue the PVC pieces and elbow.

Side-to-side Plumbing (U-turn)

Install two 3″ bulkhead fittings on the east end of the troughs, one on each side, as low as possible above the 2×3 bottom plates. Connect the two bulkheads using 3″ pipe and two elbows, keeping the pipe as close as possible to the trough endwall (to keep the walkway as open as possible). This is

Trough drain standpipe.

Trough U-turn plumbing.

the U-turn where the water changes direction and flows back towards the sump.

Repeat the plumbing installation for the other two troughs.

Estimated Parts List for Trough Construction

Here is an estimated parts list for the construction of **three 86′ troughs:**

- 1,800′ of 2×3
- 27 sheets of ½″ plywood
- 270′ of 14′ wide liner
- 54 ¼″×4″ lag screws
- 1½″ screws
- 3″ screws
- ½″ staples
- Plumbing fittings
- Paint

Note that this is an estimate. As with all construction, do your own calculations prior to ordering and purchase extra.

Raft Construction

Manufacturing rafts is a large repetitive job. For our design, you will be constructing a minimum of 258 rafts with 8,256 total holes. Treating the job like a factory line assembly and using high-quality tools is highly advised.

You must buy styrofoam that is extruded polystyrene. All other forms of styrofoam will break down and/or leach into the water. You will almost certainly need to special order your styrofoam, which will come in 2′×8′ sheets that are 2″ thick. Each sheet will be cut in half, creating two 2′×4′ rafts.

The first step is to create a template for the rafts from ½″ plywood. Rip the plywood to exactly 2′×4′. As described in Chapter 2, we use a 32-hole layout that allows for a variety of plant spacings. The hole layout is 4 across the width and 8 along the length of each raft in a checkerboard pattern. Each hole is on 6″ centers, meaning the center of each hole will be 6″ from the centers of each of its four neighbors. The centers of the holes around the outside are 3″ from the edge of the raft.

After you have marked out the hole centers, use a 2″ hole saw to complete the template. You will be using 2″ net pots which are designed to fit into 2″ holes. Use a hole saw for accuracy and a corded electric drill to cut all holes (template and styrofoam).

Using the outside edge of the template for a guide, cut each piece of styrofoam in half, creating two 2′×4′ rafts. To make the cut, either saw the styrofoam with a bread knife or deeply score it with a utility knife then snap the piece in half. Securely clamp the template to the top of a raft with a sacrificial sheet of plywood between the raft and the workbench. Drill 32 holes into the raft through the holes of the template.

You will be removing more than 8,000 pieces of styrofoam with a hole saw. This can be time-consuming and laborious. Removing styrofoam plugs from the hole saw can be particularly frustrating. The best method we developed is this: screw two 3″ screws through a scrap of plywood, 1″ apart. With the exposed screws facing upwards, screw the plywood securely onto any solid base. To clear the plug from the hole saw, push the plug down onto the screws, then twist and lift at the same time. The styrofoam should remove cleanly and remain on the screws for disposal.

Repeat the process until you have created all of your rafts.

While 258 rafts is the minimum required to fill the troughs, we suggest creating at least 10 additional spares with holes cut out and keeping an additional 50 blank rafts (without holes) for use as placeholders while harvested rafts are washed and replanted.

Painting the Rafts

The rafts must be painted white on all areas that will be exposed to direct light. This means one face (the top surface), all four edges and the inside of each of the holes. Do not paint the underside of the rafts. Use the same 100% acrylic paint (with no mold or mildew protection) that you used to paint the troughs. A "brilliant white" base is ideal. The paint must be a gloss or semi-gloss as this will be easier to clean and will stand up to regular cleaning better than flat or matte.

The styrofoam will have printing on one side; make this the top surface. Use a roller to paint the top surface and the edges. To paint the insides of the holes, dip the end of a paint roller in the paint and push it through each hole. We suggest using two rollers: one for the top and edges, one for the holes. Paint a minimum of two coats.

Painted rafts need to cure for at least one week prior to being put into the system. If the weather is compliant (not raining), the ideal place to cure the rafts is outside under the sun. After a minimum of one week, wash the rafts with a mild detergent. Don't scrub hard.

If treated with care and not scrubbed aggressively when cleaning, rafts should last many years, though they will need to be repainted every few years as dictated by the degradation of the paint due to friction, scrubbing and UV exposure.

As the rafts are unidirectional (the painted surface must always be up), the underside of the rafts will become coated with a biofilm of beneficial bacteria. Never scrub or wash the underside of your rafts. The top will be cleaned after each harvest (see Chapter 9). The bottom must be left alone or, if needed, gently sprayed clean with a hose.

When rafts are not in the troughs, always protect the unpainted sides from sunlight as the UV radiation will quickly degrade them. Degradation will be noticeable as a fine dust when the raft is wiped. Though the dust is non-toxic, it should be kept out of the system.

Source Water Installation

The installation of source water will vary from site to site. It is likely that you will need to use pex, not simple garden hose. You may be able to run the water line in the same trench as the electrical cable. Follow code.

Source water is required at a filling station over the sump, at the sink, at a hose faucet outside for washing rafts and filters and to connect the washing machine. The filling station should have at least two valved lines: one to top up the sump on a regular basis via a timer and one to fill bins for jobs such as washing plants. We recommend a tall pre-rinse faucet over the sink.

Aquaculture Subsystem Installation

This section covers installation of the fish tanks, Tank Manifold, Radial Flow Separator (RFS), Combination Filter Box (CFB), pumps, UV sterilizers and the plumbing connections between them.

Much of the plumbing of your system, the arteries of the farm, will be hidden underground. The pipes will all be near ground level, usually only a few inches below the surface. The pipes do not need to be sloped downhill in the direction of flow. Level is satisfactory. Be mindful to avoid airlocks which can happen when pipes are arranged in an "n" or "u" configuration.

The plumbing installation, both under- and aboveground, will be one of if not the most complicated and precise construction jobs. It is complex and must be done with a high degree of accuracy in three dimensions. We strongly recommend you hire a professional plumber.

IMPORTANT

Use only PVC glue that is safe for drinking water systems and use it sparingly to avoid excess glue squeezing out of the joints. Dry fit each section of plumbing prior to gluing.

All pipe in the facility is Schedule 40 PVC. Unless we specify otherwise, when we refer to pipe, we are referring to Schedule 40 PVC.

Wherever possible, use sweeps for 90° bends rather than elbows. Sweeps have approximately 30% less friction loss compared with elbows.

Follow this order for construction and installation:

1. Layout
2. Rough in the Main Waste Pipe (MWP)
3. Install the fish tanks
4. Build and install the Standpipe Assemblies (SPAs)
5. Install the Radial Flow Separator (RFS)
6. Install the Tank Manifold
7. Connect the SPAs to Tank Manifold to RFS
8. Build and install the Combination Filter Box (CFB)
9. Connect the CFB to troughs
10. Install the UV units
11. Connect UV units to fish tanks
12. Install the pumps
13. Connect the pumps to the UV units

Layout

Before breaking ground to lay pipes, you must first lay out all the major components: the four fish tanks, RFS, Tank Manifold and CFB. We suggest you also lay out the Seedling Table and workbench. Use non-toxic marking spray paint. Accuracy is important, not to the inch but within a few inches at most in all directions for each object. For round objects like the fish tanks, create accurate circles by driving a stake in the center of the object location and using a marked string of the required radius to create the circumference.

You may need to play around with the layout a bit, remarking the locations of items.

The basic requirements of the layout for both systems are:

• Sufficient space to move and work around the outside of all four tanks (should be 3′ minimum at all points)
• Access to the RFS and SPAs from the south
• The space between Tanks 2 & 3 will be aligned with the centerline of the greenhouse and the middle wall of the center trough
• The eastern edges of Tanks 2 & 3 should be 3′ from the western edges of the troughs
• Leave a 1′ gap between Tanks 1 & 2 and 2 & 3 at their closest points

Walk around the layout you have created. Imagine the height of the objects and confirm that you are happy with both walking and working space.

Once you are satisfied with the spray paint layout, mark the location of the four Standpipe Assemblies. The SPAs for Tanks 1, 2 & 3 will connect to both the Tank Manifold and the Main Waste Pipe (MWP). The SPA for the Purge Tank will connect only to the MWP unless it is being used as a hatchery tank or for a 4th cohort, in which case it will also connect to the Tank Manifold.

Installing the Main Waste Pipe

The first pipe to lay is the MWP from the RFS towards the west end of the greenhouse. The MWP is a 4″ pipe that carries effluent from the RFS and the four fish tanks to the Waste Tanks outside the greenhouse on the western side.

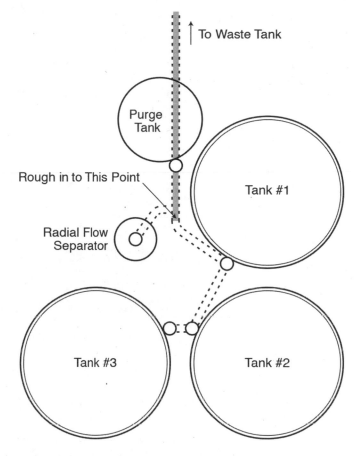

Main Waste Pipe plumbing.

Dig a trench from the Waste Tank that runs under the sliding doors and ends at the rough-in point between the Purge Tank, RFS and Tank 1. The top of the pipe at this point should be 6–12″ below ground.

Install the primary Waste Tank in its excavated site at the appropriate height.

Glue the pipe sections together, lay them in the trench and install a temporary cap to prevent dirt from entering the pipe. Backfill most of the trench, leaving the capped end exposed. You will later connect this to the RFS and the four SPAs.

Installing the Fish Tanks

The first step is to build a support ring on the ground for each tank. The blocks you use must be capable of forming a circle. For this we used and recommend Allan Block retaining wall blocks as they have wings that can be easily removed, converting each block into a wedge shape which is ideal for forming circles. The support ring will be filled with a bed of sand, and the tanks will sit approximately halfway onto the blocks. The tanks will not sit directly on the blocks, they will only contact sand.

At the location of the four SPAs, leave a gap in the support ring wide enough for each tank's bottom drain pipe to pass through (approximately 6″). Starting with the gap, work your way around the circumference of each ring, setting the blocks into place. Make a bed of sand for each block as you go. Use a hammer and wooden striking block to firmly set each block into the sand. Use a string line from a peg in the center of the circle to ensure it is perfectly round. Use a surveyor's level to ensure that each block is installed at the same height. Use a builder's level to confirm the height.

IMPORTANT

It is critical that the top edges of the tanks are all the same height in relation to one another.

Tanks 2 & 3 on bases with roughed-in plumbing and RFS.

Each ring (and the hollow core of each block) is filled above the lip of the blocks with fine bedding sand. It is very important that there be no rocks or sharp objects in the sand. Compact the sand with a hand tamper or small plate compactor.

IMPORTANT

It is crucial that the sand installation be sufficient so that the tanks only contact sand. Expect some settling when placing the tanks due to the pressure of tons of water.

The bottom drain is located at the bottom of the each tank in the center. Dig out a depression in the middle of the ring appropriate to the tank design and a trench for the drain pipe from the center of the circle through the gap in the blocks. Many tanks will have a gentle slope in the bottom. Try to match the profile of the base of the tank with the sand. Lightly wet the sand prior to digging to create more precise shapes.

Tanks will typically be shipped upside down. While upside down, glue a pipe into the drain of the tank. The pipe will not be cut to final length at this time and should simply extend past the blocks by a foot or two. This will later connect to a tee at the bottom of its SPA.

Once the glue is set, carefully turn the tank over with a team of at least four people and gently set it into place. Check that it is level using a straight 2×4 and a builder's level. The tank may need to be shifted around slightly to settle it into a level position.

Repeat the process for the other three tanks. Use a surveyor's level to ensure the tops of all tanks are level with each other.

Building the Standpipe Assemblies (SPAs)

The SPAs for Tanks 1, 2 & 3 each have one inlet (from the tank) and two outlets: to the Tank Manifold and to the MWP. The SPA for the Purge Tank is the same construction, only a smaller diameter, and the outlet is not connected to the Tank Manifold unless it is going to be used as an additional culture tank.

The only difference between the SPAs for Tanks 1, 2 & 3 is that the SPA for Tank 3 has an elbow at the bottom as it is the beginning of the MWP, whereas the SPAs for Tanks 1 & 2 (and the Purge Tank) have tees where they connect inline to the MWP.

Each of the SPAs for Tanks 1, 2 & 3 will have a 4″ outer pipe, a 2″ inner pipe, and a 4″×2″ bushing at the bottom. The 2″ pipe runs inside the 4″ pipe and plugs into the bushing to block the lower outlet to the MWP. Lifting the inner pipe will allow effluent to flush into the MWP and out to the Waste Tanks. The 2″ pipe extends approximately 12″ past the top of the 4″ pipe so that it can be easily gripped for lifting. The 2″ pipe must be capped to prevent odor leakage from the Waste Tanks.

The SPA for the Purge Tank is built in the same manner except that the outer pipe and both tees are 3″, the inner pipe is 1½″, and the bushing is 3″×1½″.

The construction of the SPAs is straightforward. Simply follow the diagram. Start by modifying a 4″×2″ bushing by cutting off the wrenching fins to allow it to fit backwards into the 4″ tee. The fins must be cleanly removed without taking off too much material. There must be no gaps between the bushing and the tee. Use an oscillating tool or fine tooth hacksaw to remove the fins, then sandpaper to round off the bumps until the bushing fits snuggly into the tee.

Standpipe Assemblies. Note that each SPA sets its tank water height.

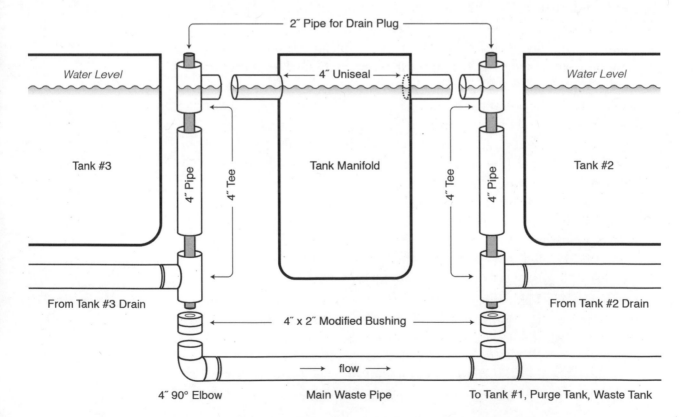

Glue the bushing in reverse exactly halfway into the 4″ tee. The other half of the bushing will be connected to the MWP.

At this time, install a 4′ length of 4″ pipe in the top of the lower tee. You will install the upper tee that determines the water height later.

IMPORTANT

The upper tee will not be glued so that you can swivel it or correct its height as necessary.

Connect each SPA to its tank drain as shown in the diagram. If necessary, use a coupling and a length of pipe to extend the tank drain pipe to the location of the SPA .

Connect the Purge Tank SPA to the MWP via a tee in the MWP, then extend the MWP to connect to the SPAs of Tanks 1, 2 & 3 in that order.

Rough in the plumbing from the RFS to the MWP. The connection point will be between the Purge Tank and Tank 1. Use pipe sized for the outlet of the RFS. Cap the pipe to prevent debris from getting in.

The height of the upper tee in each SPA determines the height of the water in the corresponding tank. Finalize the height of the SPAs by cutting the vertical 4″ pipe and fitting the upper tee on each. The top of the outlet of the tee must be 3″ below the lip of each tank. It is vital that this be exact as this will determine the height of the water in the tank.

Installing the Radial Flow Separator

The RFS has one inlet that comes from the Tank Manifold and two outlets: a lower outlet that connects to the MWP to remove effluent and an upper outlet that connects to the CFB.

Once in its final position, the top of the RFS must be 4–6″ lower than the bottom of the outlet pipes on the 4″ upper tee of the SPAs. The RFS may therefore need to be partially buried. Use a surveyor's level to set the RFS at the correct height, making sure it is perfectly level and that the upper outlet is pointed in the direction of the CFB. Use a compacted gravel or cement block base to prevent the RFS from settling or tipping. Connect the lower outlet of the RFS to the MWP and backfill around the base of the RFS.

Once the RFS and all SPAs have been set in place and connected to the MWP, the MWP and the bases of the SPAs can be backfilled.

Installing the Tank Manifold

The Tank Manifold must be a food-grade plastic barrel or container, ideally with a lid. A 50-gallon polyethylene barrel works well as it has sturdy walls that can be easily cut or drilled and a wide circumference which is helpful for attaching fittings.

The top of the Tank Manifold is at least the same height as the tops of the fish tanks, and the bottom is 6–8″ lower than the bottom of the RFS inlet, leaving just enough room for a 4″ bulkhead. If the bottom of the Tank Manifold is lower than this, large settleable solids from the fish tanks will collect here instead of being carried into the RFS, and the Tank Manifold will need to be cleaned out regularly. If needed, build a small platform similar to the tank rings to elevate the Tank Manifold to the correct height. If the top of the manifold is significantly higher than the tanks, it can be cut down to a few inches above but no lower than the top of the fish tanks.

Once the Tank Manifold has been set in place at the correct height, install three 4″ Uniseals facing towards the SPAs of Tanks 1, 2 & 3 at the same height as the outlets of the upper tees in the SPAs. Connect the SPAs to the Tank Manifold with lengths of 4″ pipe. Use 45° or 22.5° fittings as needed.

Connecting the Tank Manifold to the Radial Flow Separator

The inlet on the RFS will connect to the outlet on the Tank Manifold. To create the outlet on the Tank Manifold, install a Uniseal on the manifold at the same height at the inlet on the RFS. The Uniseal and connection pipe size will be the same as the inlet on the RFS.

Constructing the Combination Filter Box (CFB)

The CFB is a custom filtration component that must be well-built precisely to the following specifications.

We recommend building the CFB from ¾″ marine-grade plywood, waterproofed with epoxy resin. If different materials are used, follow the same design guidelines and use the exact dimensions as outlined.

Use the dimensions shown to cut the bottom, lid, sides and ends of the box from two sheets of plywood. The inside dimensions of the finished box are 58.5″ long by 24″ wide by 20.5″ tall. The ends are sandwiched between the sides.

Glue and screw the sides and ends together using 2″ exterior screws and epoxy resin, then glue and screw the bottom onto the box. Make sure it is aligned squarely. Ensure that all joints are tight with no gaps. Make sure to apply enough epoxy resin that it squeezes out of the joints, then wipe smooth the excess.

To make the ribs, cut 24 strips of ¾″ plywood that are 1″ by 19¾″. Install the ribs according to the diagram, with 9 pairs of ribs at the upstream side (inlet from RFS) and 3 pairs at the downstream side (outlet to troughs). The ribs are 1½″ apart to house the filter screens. The space in between the two groups of ribs is for the Moving Bed Bio-Reactor (MBBR). There will be 3½″ between the outer ribs and the ends of the box.

Note that ribs are required on both (opposite) sides. The diagram shows ribs only on one side of the CFB in order to show dimensions clearly.

The ribs are flush against the bottom of the box with a ¾″ gap at the top. Drill pilot holes then fasten the ribs to the inside of the box using 1¼″ exterior screws. Screw from the outside of the box into the edge of each strip. The ribs stand 1″ in from the walls of the box. Ensure the strips are snug against the inside.

The lid is fitted with four strips of ¾″

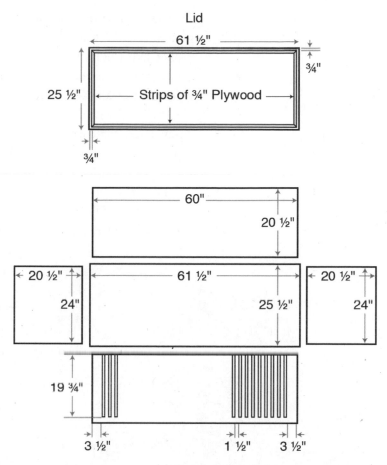

CFB construction blowup. Note that the ribs, shown here only on one side, are installed on both sides.

plywood on the underside to hold it in place. The outer edges of the strips are ⅞″ in from the edge of the lid. When in place, the strips fit snugly inside the walls of the box, on top of the ribs, and the lid's weight keeps the filter screens tight against the bottom of the box.

Cut two holes for bulkhead fittings, one at each end of the box. Match the inlet bulkhead in the north end to the RFS outlet and cut as high as possible without interfering with the lid. The outlet bulkhead on the south end (to the troughs) is 4″, and the top of the hole is 6″ below the top of the box wall.

Cut a hole for the 1″ bulkhead for aeration in the east side in the center of the area for the MBBR. The top of the 1″ bulkhead is 1″ below the top of the side.

Beside the inlet from the RFS in the upstream (north) end, cut a hole for the 1½″ bulkhead for the bypass from the pumps.

Cut each hole with a hole saw that is just large enough for the bulkheads to fit snugly through. Test fit the bulkheads but do not install them now.

The CFB is now assembled, time to reinforce and waterproof it with fiberglass and epoxy resin.

See IMPORTANT note (page 80) on epoxy and fish safety.

Reinforce the four vertical corners of the box by filleting the seams with wood and/or a bead of thickened epoxy resin. A "fillet" coves and strengthens the inside corners where two plywood panels meet. Before the fillet fully cures, cover the corner with a 6″ wide strip of fiberglass cloth. Immediately saturate the cloth with resin and push out all the air bubbles with a plastic scraper or fiberglass roller. Allow the resin to cure until it is firm but still soft enough to indent with a thumbnail.

Fiberglass corner fillet in the CFB.

Before the fillet is fully cured (typically 1–3 hours), waterproof the box with resin. Saturate the entire inside with three coats. Successive coats can be added once the previous coat has "kicked off," or gelled over, but before it is fully cured. If the epoxy becomes too hard to indent with a thumbnail between coats, wait until it is fully cured before recoating.

Be sure to fully saturate all seams, corners and around the holes and top edges of the box. Be generous with the resin — the plywood will soak up considerable quantities, particularly the edges.

Saturate the sides and edges of the lid and the lid strips with three coats of resin. The lid will inevitably become valuable bench space, so protect it well. Saturate the outside of the box with two coats of resin.

Allow the resin to cure for at least 24 hours. After curing, there will be a waxy "amine blush" left on the surface of the epoxy. This residue is water-soluble and must be thoroughly washed off using only warm water and a soft cloth. Do not use any solvents or detergents.

Use 120 grit paper to lightly sand the gloss off the epoxy on the outside of the box (all four sides and lid), then paint it using the same white paint used on the rafts. Do not paint the inside of the box or the bottom of the lid.

Attach the bulkhead fittings to the box, clamping the gaskets tightly inside.

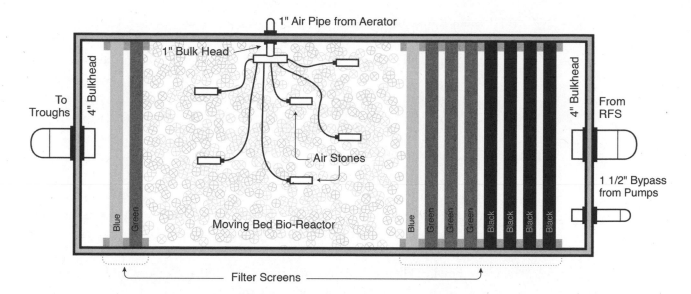

1" Air Pipe from Aerator

1" Bulk Head

4" Bulkhead

To Troughs

Air Stones

Moving Bed Bio-Reactor

Blue
Green

Blue
Green
Green
Green
Black
Black
Black
Black

4" Bulkhead

From RFS

1 1/2" Bypass from Pumps

Filter Screens

CFB top view.

Installing the Moving Bed Bio-Reactor (MBBR)

On the inside of the 1″ bulkhead, assemble the aeration system for the MBBR. This consists of a check valve, a tee, two 6″ arms with caps, six ¼″ MNPT x barb fittings, ¼″ air tubing and six air stones.

Cut two 6″ lengths of pipe to create a manifold with two arms. In each arm, drill three holes that are slightly smaller than the threaded end of the ¼″ MNPT x barb fittings. Use a ¼″ thread tap to cut threads into the holes and screw in the barbs using Teflon tape. Cap the arms.

Inside the CFB, install a low-pressure swing check valve on the 1″ bulkhead to prevent water from backflowing down the air pipe. Install a 1″ tee onto the check valve and glue the two arms to the tee. Cut six lengths of ¼″ air tubing to equally space the air stones around the bottom of the MBBR. Attach the air stones to the barb fittings in the arms.

IMPORTANT

Install the ¼″ MNPT x barb fittings prior to gluing the arms into place.

Installing the Filter Screens

Each screen is 24″ wide by 19.75″ tall and 1.5″ thick. We use and recommend Matala filter media which comes in 48″ by 39″ sheets, 1.5″ thick. Each sheet is cut down the middle both ways to create four pieces that fit perfectly into the CFB.

Matala filters are available in four grades, identified by different colors. Starting from the inlet side, use four black (low-density) screens, followed by three green (mid-density) screens, then one blue (high-density) screen. On the other side of the MBBR, install one green then one blue. These last two screens act as a containment barrier for the MBBR media, preventing it from flowing out of the box.

The CFB is specifically sized to house Matala filters with no excess. If you choose to use another brand or style of screen in your box, you may reconfigure the dimensions of the box. If you do so, note that the total volume must not decrease and the overall shape and layout should be as close as possible to our design.

Finally, fill the MBBR with the filter media. We use and recommend Sweetwater SWX bio-media which is sold by the cubic foot. You will need four cubic feet which will provide about 93 m² (1000 ft²) of BSA.

Filter screen (right) and MBBR media (left) in the CFB.

Installing the Combination Filter Box

Position the CFB as close as possible to the outlet of the RFS, allowing room for plumbing connections and to comfortably walk and work around it. Also consider that in-ground plumbing will be laid close by.

The height of the CFB is determined by the height of the outlet pipe on the RFS. Typically the RFS outlet will be pointing down, so you will need to fit an elbow on the outlet and set the CFB high enough that its inlet lines up with the elbow.

Mark the final placement and build a base for it out of cement blocks, using a surveyor's level to determine the correct height. Double-check that the box is perfectly level in all directions and is at the correct height before making plumbing connections.

The upper outlet on the RFS connects to the inlet on the CFB. You have already cut the CFB inlet to match the size of the RFS outlet and installed a bulkhead. As with all plumbing in your system, use as few bends as possible and use sweeps instead of elbows where possible.

Installing the Underground Plumbing (Trough Side)

The CFB will connect underground to the troughs. The pipe will run between the sump and Tank 3. For a tower system, the CFB will connect to the northeast corner of the sump.

Dig a trench that is 6″ deep and 12″ wide along the west end of the troughs and between the south trough and the sump. It should be as close as possible to the trough endwalls and run from the south end of the south trough until it is in line with the aerator. The trench will contain the trough drain pipe, the trough inlet pipe and the air distribution pipe.

Dig a second trench between the sump and Tank 3 that is 6″ deep and 6″ wide and that runs from the outlet of the CFB east to intersect with the first trench. This will contain the 4″ pipe running from the CFB to the trough inlets.

Lay the pipes in the trench. The 4″ trough drain pipe is closest to the troughs (farthest east in the greenhouse). The 4″ inlet pipe lies beside it to the west. The 2″ air pipe sits in the groove between the two 4″ pipes.

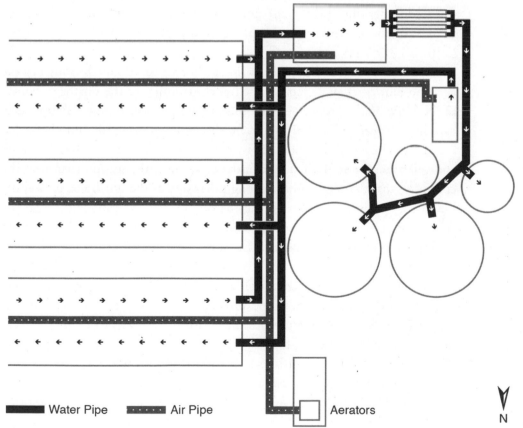

Water Pipe ▪▪▪▪ **Air Pipe**   **Aerators**

Water and aeration plumbing schematic. Note that water plumbing from tanks to Tank Manifold to RFS to CFB is not shown.

Lay the 4″ trough drain pipe from the outlet of the north trough to the center of the sump. The pipe must be at the right height to pass through the sump collar. Drill a hole in the collar, as low as possible and just wide enough for the pipe to fit tightly. Before inserting the pipe, seal the inside of the hole from moisture with epoxy resin. Inside the sump, place a 4″ elbow on the end of the pipe. Do not glue the elbow.

Connect the trough drain pipe to the outlets on the south side of each trough. Cut the pipe below the trough outlets and install a 4″×3″ tee. Using two short sections of 3″ pipe and one elbow, connect each trough outlet to the trough drain pipe. Keep everything as close to the troughs as possible.

Using 4″×2″ tees, connect the 4″ trough inlet pipe to the valves previously installed on the north side of each trough in the same manner as the drains. Where the inlet pipe intersects the trench from the CFB, use one sweep to continue the inlet pipe towards the CFB. As close to the CFB as possible, connect the underground pipe to the outlet of the CFB using sweeps.

Installing the Aeration Pipe

The 2″ underground air distribution pipe runs from the aeration blowers (located under the workbench) along the west end of the troughs with one branch to the northeast corner of the sump and another branch to the CFB.

Extend the existing trough drain/inlet trench past the north trough to the aeration blowers. Lay 2″ pipe in the trench, running to the northeast corner of the sump, in the groove between (on top of) the two 4″ pipes. Temporarily cap the end near the aerator to prevent dirt from entering (do not glue).

At the sump end, use one elbow to run the pipe from the south trough endwall to the sump. Drill a hole in the east side of the sump collar, near the corner, so that the pipe will run along the inside of the north collar. Seal the hole with epoxy resin before fitting the pipe. Once the pipe is through the collar, glue a 2″ ball valve onto the end. Later, you will construct a manifold to distribute air to the diffusers in the sump. See diagram showing the sump top view on page 79.

Cut the air pipe at the center of the middle wall of each trough and install a 2″×1″ tee. Place a 1″ riser pipe in each tee so that it rises a few inches higher than the endwall, directly in line with the center of the middle wall. Temporarily cap the risers (do not glue).

Cut the air pipe where it intersects the trench to the CFB, beside the 4″ inlet pipe. Insert a 2″ tee and run the air pipe to just below the 1″ bulkhead on the east side of the CFB. Use a 2″ elbow, a 2″×1″ bushing, 1″ pipe and another 1″ elbow to connect the air pipe to the 1″ bulkhead on the CFB.

Once all pipe is glued and installed, backfill the trenches then add a layer of crushed gravel to make a flat surface.

Aeration Blowers Installation

IMPORTANT

To prevent the possibility of water siphoning into the aerators, the outlet pipes must be installed higher than the surface level of the water in the troughs. You will need to build a platform for the aerators under the workbench.

Aeration blower in sound-dampening box with front of box removed.

As discussed in Chapter 2, our design calls for installation of two identical aerators for redundancy. The aerators will be located under the workbench, next to the north wall of the greenhouse. Follow the manufacturer's instructions closely.

Install the aerators in parallel, with their outlets connecting to a 2″ tee on the end of the underground distribution pipe. Install a ball valve on each aerator before the manifold so that you can easily switch between the two. Only one aerator will be active at any time. Optionally, install an air pressure switch and a relay switch between the two aerators so that if one fails, the other automatically turns on. If installing a relay switch, you must use swing check valves rather than ball valves.

If you only install one aerator (which we do not recommend), you must connect the aerator to the monitoring system to both inform you of a unit failure and automatically activate the backup oxygen system. See Chapter 5 for details on backup oxygen and monitoring system.

An aerator can be quite loud, and it will be operating 24/7 in your greenhouse. A simple sound-dampening box constructed of lumber and plywood, lined with rigid styrofoam and/or carpet will greatly reduce the noise. It is vital, if building a box that it NOT be sealed. Air must be allowed to freely enter the box both for supply to the unit and for the unit to remain cool. Our recommended method is to leave a 2″ opening across the width of the box at the top of the front and back.

If your aerators come with a length of rubber pipe to connect the outlet of the aerator to the air distribution line, use it. This rubber pipe is designed for dissipating any heat that may build up from the compression of air through the blower. In extreme cases, if the air pressure in the pipe is high enough, the temperature can get hot enough to melt PVC. Our system is designed to operate at less than 1 psi, so heat should not be an issue.

Like a water pump, an aerator is described by a performance curve graph that shows the flow rate, measured in cubic feet per minute (CFM), at a given depth of water. Unlike a water pump, the friction losses associated with pumping air are miniscule, and total head does not need to be calculated. The only substantive restriction to flow is the depth of water at which the air leaves the diffuser.

The size of the aerator is determined by the number of air diffusers in the system multiplied by the manufacturer's suggested flow rate. We recommend using one Sweetwater AS5 air stone under every raft. The AS5 stones are 3″×1″ and have a suggested flow rate of 0.3 CFM each. For 86′ troughs, you will require 258 airstones, which equal 77.4 CFM (258×0.3).

For the sump and the MBBR, we recommend using sixteen Sweetwater AS15 air stones (ten in the sump, six in the MBBR), with a suggested flow rate of 0.5 CFM each for a total of 8 CFM (16×0.5).

The total airflow required for 86′ troughs is therefore 85.4 CFM @ 10″ of water. We recommend the Sweetwater SST30 regenerative blower (or comparable). The SST30 is slightly bigger than is required, but it is better to oversize the aerator. It is not possible to over-oxygenate the system, and under-oxygenating is detrimental to all living things and may lead to system crashes.

IMPORTANT

The goal is to fully saturate the water with oxygen by the time it exits the sump so that the fish receive as much oxygen as the water can hold.

Once the aerators have been installed under the workbench, connect them to the underground air distribution pipe.

Install a dedicated power circuit for the aerators (see Electrical Installation, this chapter).

Air Stone Installation

IMPORTANT

Drill and install the ¼″ MNPT x barb fittings in the 1″ air pipe prior to installation of the air pipes on the middle wall (see below for instructions).

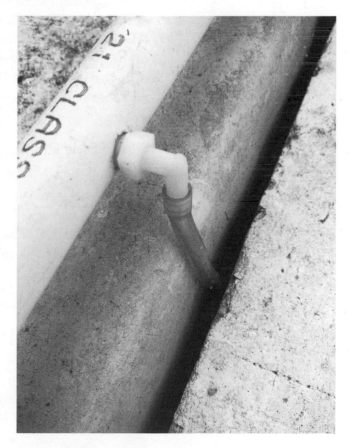

Air is delivered to the stones in each trough by a 1″ distribution pipe running along the top of each middle wall. Air stones are located under every raft (every 2′) and connected to the distribution pipe via a ¼″ 90° MNPT x barb fitting and 3′ of ¼″ vinyl air tubing. The fittings are elbows and have a threaded end and a barbed end. The threaded end connects into the 1″ pipes, and the barbed end connects to the ¼″ air tube.

Each 86′ long trough requires 86′ of 1″ pvc pipe, 86 Sweetwater AS5 air diffusers (43 per side), 86 MNPT x barb fittings, and 258′ of ¼″ vinyl tubing.

Lay out the 1″ pipe and mark the locations of the diffuser connections. Make two marks 1′ from the end of the trough and then two marks every 2′. The two marks are on opposite sides of the pipe (180° from each other, facing towards the rafts).

To install the fittings, drill ³⁄₁₆″ holes in the 1″ pipe at each mark. Drill slowly and without applying too much pressure, or you will crack the PVC. Cracked or otherwise ruined PVC must not be used. The spoiled sections can be cut out and coupled back together if desired. Expect to spoil some pipe as this is a finicky job, which is why we recommend installing the fittings prior to installing the pipes on the middle walls.

¼″ 90° MNPT x barb fitting for aeration.

Once you have drilled the holes, use a ¼″ thread tap to create threads on the inside of the holes. Thread a fitting into each hole using Teflon tape to keep it airtight. Complete all fitting installations, then glue the distribution pipes on the middle walls of each trough pair.

At each trough, remove the temporary cap on the 1″ riser. Connect a 1″ elbow to the top of the riser and lay 1″ pipe with fittings installed along the length of the top of the middle wall. Cap the air pipes at the eastern ends of the troughs. Use pipe straps every 10′ to secure the air pipe to the top of the middle walls. It is acceptable to penetrate the top of the LDPE only for this purpose. Remember: do not wear shoes in the troughs.

Once the air pipes have been installed, connect 3′ of ¼″ vinyl air tubing to the barbed end of each fitting. Connect an air stone to the other end of tube and fit two flexible O-rings over each air stone. The O-rings will act as bumpers to prevent the rough stones from rubbing on the trough liner.

In the sump, aeration is provided by a 2″ pipe that runs along the inside of the north collar, entering from the east. Install eleven ¼″ MNPT x barb fittings: ten for air stones and one for an air pressure switch which connects to the monitoring system. Connect air tubing to each fitting and air stones with O-rings on ten of them. Run the air tubing for the monitoring system to the west end of the sump, where it will later be connected to an air pressure switch.

Pump Side Plumbing Installation

Installing the UV Sterilizers

The first step is to determine the units you will be installing using the information in Chapter 2. The layout of the units will depend on the number you will be installing. The recommended units (Smart UV High-Output E150S) are 5′ long each. All units must run parallel to each other on a horizontal plane and should be as low as possible to the ground to minimize system head.

IMPORTANT

If for any reason the main pumps need to be temporarily shut down, also shut down the power to the UV units. Leaving the bulbs powered with no water flow will rapidly increase the water temperature in the UV units which can degrade the life of the bulb.

The units are located between the sump and the southwest corner of the greenhouse, running east-west. Support the sterilizers by building a rectangular frame out of 2×12 lumber and cutting semicircles in the east and west ends to seat the round bodies of the units. The frame must be tall enough to accommodate the connections from the pumps and the replacement of the bulbs.

Using a jigsaw, cut semi-circles the same diameter as the unit bodies so they will fit snugly. The sterilizers must be evenly spaced and at the same height. Secure the units to the frame with galvanized or stainless steel strapping.

The UV bulbs will be almost the same length as the housing and are replaced by sliding in and out of one end of the housing. The units must be high enough and installed in the correct orientation to allow enough room to change the bulbs above the lid of the sump.

Due to the design of the UV units, they are prone to trapping bubbles of air in the exposure chamber between the quartz sleeve and the sterilizer body. To prevent this, install the inlet and outlet pipes facing up and install the units on a slight uphill slope of no more than 1″ from end to end. To do this, shim the 2×12 frame so that the inlet (east) ends of the sterilizers are lower than the outlet (west) ends. Installing the units in this manner will prevent airlocks from forming as any trapped air will be pushed through by the flow of water and released through the fish tanks.

Main water pumps and UV sterilizers. Side view.

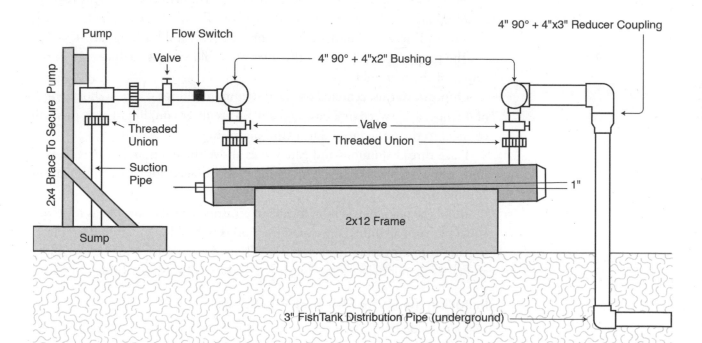

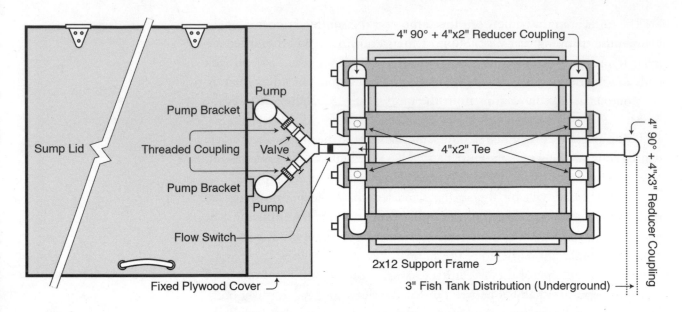

Main water pumps and UV sterilizers. Top view.

Once the frame is in position with the units strapped into the semicircles with outlets facing up, construct the two manifolds that will connect the pumps to the sterilizers and the sterilizers to the fish tanks.

Build both manifolds out of 4″ pipe, tees and elbows. The two inner UV units will connect to the manifolds via 4″ reducer tees, and the two outer units connect via 4″ elbows and reducer bushings. The reducer tees, and reducer bushings are 4″×(X)″ according to the size of UV inlets (e.g. if the UV sterilizers have 2″ inlet and outlet connections, the tees and bushings are 4″×2″).

The pumps will connect to the inlet manifold via a 4″ reducer tee in the center of the manifold.

Opposite to this, centered on the outlet manifold is a 4″ tee, a short length of 4″ pipe, a 4″ sweep or elbow and a 4″×3″ reducer coupling. Later, this will connect to the underground fish tank plumbing.

Place threaded unions and gate valves above the inlets and outlets of each unit. Many sterilizers come with unions pre-installed for ease of installation and maintenance.

Build the manifolds symmetrically to ensure the water flow rate is relatively equal between all units. The easiest method is to build the manifolds in place, starting with the connections to the sterilizers. Dry fit everything before gluing and take care not to let excess glue drip into the sterilizers, gate valves or unions.

Installing the UV to Fish Tank Plumbing

Dig a 6″ wide by 6″ deep trench from the UV outlet manifold to the Purge Tank SPA (just to the west of the CFB). Continue the trench to the SPAs of

Tanks 1, 2 & 3 in that order. Minimize the system head by softening bends or using sweeps wherever possible. See diagram showing DWC plumbing on page 105.

Use 3″ pipe to run underground from the UV outlet manifold to Tank 3.

Just past the northwest corner of the CFB, install a 3″×1½″ tee, 1½″ riser pipe, 1½″ ball valve and 1½″ elbow to connect to the 1½″ bulkhead fitting in the CFB. This bypass will help to balance the flow and allow the tanks to be taken offline without shutting off flow to the CFB and troughs.

Install a 3″×1½″ tee next to the SPA location for the Purge Tank and Tanks 1 & 2 (three tees total). Next to the SPA location for Tank 3, install a sweep and a 3″×1½″ bushing. Install four 1½″ riser pipes in the reducer tees and reducer bushing, extending up to the lips of the tanks. The risers must rise to touch the outer edges of the tank lips so that they can be secured to the lips with a wood backing block or metal L-bracket.

Cut the riser pipe 12–16″ below the bracket. Install a 1½″ ball valve and 1½″ threaded union between the valve and the bracket. Using short lengths of pipe and an elbow, make a spout and fit it in the top of the threaded union. The spout should extend 6–12″ into the tank.

Tank inlet spout.

Once everything is glued together, loosely secure the riser to the L-bracket with galvanized or stainless steel strapping. Keep this loose so that the inlet spout can be swivelled when the union is loosened.

Backfill the trench and finish the ground with a layer of compacted crush gravel.

Calculating the System Head

Now that the tank inlet plumbing is complete, you can accurately calculate the total head of the system and use that data to find a pump that matches the performance requirements of the system. Remember you will be buying two identical pumps, each sized to run the system on its own.

Calculating the total head of this type of pumping system is a complex and challenging task. The pumping system has multiple parallel branches, with different sizes of pipes and fittings, and different flow rates through the various branches (e.g., the 3″ main distribution line may

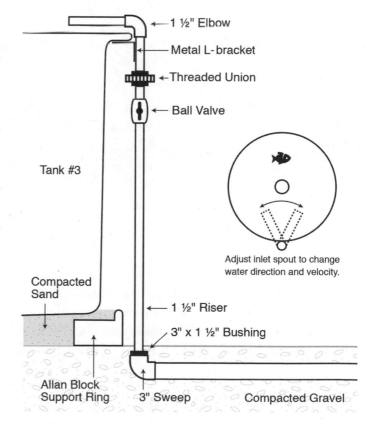

← 1 ½" Elbow

Metal L-bracket

← Threaded Union

← Ball Valve

Tank #3

Adjust inlet spout to change water direction and velocity.

Compacted Sand

← 1 ½" Riser

3" x 1 ½" Bushing

Allan Block Support Ring   3" Sweep   Compacted Gravel have a flow rate 400 LPM, but the 1½″ inlets for the tanks only have ⅓ of that). The friction losses of each branch need to be calculated separately and some, such as the friction losses from the UV sterilizers, may not be published by the manufacturer and can only be estimated. We recommend hiring a professional.

IMPORTANT

The total head for your system is calculated only on the pump side of the plumbing: from the pumps to the four tanks (not including the bypass to the CFB). The gravity side of the plumbing (from the tanks through the filter units to the troughs and back to the sump) is not included in the total head calculation.

The basic steps to determine the total head of the system are as follows:

1. For each branch or section of the system, determine the friction losses of each fitting in "equivalent feet of pipe."
2. Measure the total length of pipe used in each branch or section and add to the "equivalent feet of pipe" for that branch.
3. Use the following charts (or as provided by the manufacturer) to determine the head, in decimal feet, of each branch at the flow rate expected for that branch.
4. Measure, in decimal feet, the vertical distance between the predicted water level in the sump and the suction inlet height of the pump. This is the suction head.
5. Measure, in decimal feet, the difference in height between the pump outlet and the fish tank inlets. This is the vertical head.
6. Add the friction losses of each branch or section, the suction head and the vertical head. This is the total system head.

For example, the inlet assembly on Tank 1 has a total pipe length (including equivalent length of fittings) of approximately 18′ of 1½″ schedule 40 PVC. At a flow rate of 140 LPM, the friction loss for this branch would be 1.3′ of head.

Ideally this process should be repeated for a range of flow rates (as flow increases, so does head) and the resulting data plotted into a graph. This creates a "system curve" which is just like a pump curve except the plotted line will curve in the opposite direction.

Friction losses for schedule 40 PVC pipe, shown as "feet of head per 100′ of pipe."

Flow \ Size	1″	1½″	2″	3″	4″
50 LPM	12.0	1.5	0.45	0.07	
100 LPM		4.0	1.17	0.17	0.05
150 LPM		9.5	2.7	0.41	0.11
200 LPM		14.0	4.2	0.6	0.16
300 LPM			10.0	1.5	0.4
400 LPM			15.5	2.25	0.6

Friction losses for common PVC fittings, shown as "equivalent length of straight pipe."

Fitting \ Size	1″	1½″	2″	3″	4″
90° elbow	2.6	4.0	5.5	7.7	10.1
90° sweep	1.7	2.7	4.3	6.3	8.3
45°	1.4	2.1	2.8	4.1	5.4
Tee-run	1.7	2.7	4.3	6.3	8.3
Tee-branch	6.0	8.1	12.0	16.3	22.1

Selecting the Pumps

Once the the total system head is calculated and a system curve created, the data can be used to source a pump by comparing the performance graph of the system with that of the pump: the point where the system curve and pump curve intersect is the operating point of the system. To select a pump, look for one that can create the flow of water needed at the total head of the system plus 10%. For example, if the total head is 15.5 feet, and the optimum flow is 400 LPM, look for a pump that will output 440 LPM @ 15.5 feet.

As a rule of thumb, it is better to slightly oversize the pump than undersize it, within reason. The pump is going to be on 24 hours a day so it should be as efficient as possible. Excess flow can be routed through the bypass valve in the CFB.

There are many brands and models of pumps. Compare as many as possible to assess their performance and electrical consumption. Look for pumps that will consume the least amount of electricity and still do the job. Variable-speed pumps, although usually more expensive than single-speed pumps, can be very efficient and can adapt to a range of flow rates and conditions.

Most pumps used for similar applications are centrifugal pumps with end suction and top discharge, and many of these can be mounted vertically. Other types of pump have strainer baskets or primer pots connected to them and can only be mounted in one direction. If your chosen pump cannot be mounted vertically (so that its suction pipe points straight down into the sump), additional plumbing may be required to connect the pump to the UV inlet manifold. This will add to the total system head. If possible, choose pumps that have end suction and top discharge.

We have used Little Giant OPWG series pumps running continuously for more than four years with no problem of any kind. They are single-speed, low-head, high-volume pumps that are electrically efficient and very reliable. We recommend them, though there are many other quality manufacturers.

As discussed in Chapter 2, two pumps should be installed for redundancy. Both pumps will be connected to the system, but only one will be operating at any time. The pumps will connect to the UV inlet manifold with a wye. If it is not possible to source a symmetrical wye fitting, a tee is acceptable, but the backup pump should be connected to the branch of the fitting and the primary pump to the straight run.

Construct a mount or bracket out of 2×4 between the double joist and the western end of the sump collar to hold the pumps in position. Once the pumps are mounted and their suction pipes in place, cut a plywood cover to fit permanently over the western-most section of the sump, with holes for the suction pipes. Waterproof the cover with at least two coats of epoxy resin, then optionally paint it.

Pumps should always have threaded unions on both suction and outlet pipes to allow easy removal or cleaning of the pump. They should also have a gate valve on the discharge pipe and a swing check valve on the bottom of the suction pipe, 4–6″ above the bottom of the sump.

Place a flow switch between the pumps and the UV inlet manifold. The switch will be connected to the monitoring system later. Use either a paddle-style switch like the Aqualarm flow switch or a flow-through style like the Pentair ST9. We use and recommend the Aqualarm flow switch which has MNPT threads on the bottom that can be drilled and tapped into an existing pipe or fit into a tee with a threaded bushing of the correct size.

Heat Pump Installation

The heat pump must be sized and installed by an HVAC professional to ensure accuracy and functionality.

The heat pump must be mounted to a solid level base located just outside the south side of the greenhouse near the sump. If the unit doesn't come on a stand, pour a minimum 4″ thick cement pad and use anchor bolts to secure it. It must have a dedicated electrical circuit.

Heat pumps produce a lot of condensate, so plan for drainage. Units typically have a drain port on the bottom to attach a hose.

The heat pump will require an external pump to draw water from the sump and circulate it through the unit. The min/max flow rate is listed for each heat pump. Before buying a circulation pump, plan all plumbing according to the following directions and calculate the total head of this subsystem.

A water filter must be used on the inlet of the heat pump to avoid plugging the heat exchanger with solids from the system. We use a 200 micron disc filter which allows high flow with minimal restriction.

Place the heat pump as close to the sump as possible to minimize heat gain/loss through the plumbing and to increase pumping efficiency (less head). Once it is solidly in place, dig a trench between the unit and the sump for the water in/out pipes. Pipe sizing depends on the circulation pump discharge and heat pump connections. Make the trench big enough for both the heat pump pipes and the cistern pipes (see Cistern Installation below). Install both at the same time if possible. Make the trench wider than necessary between the sump the greenhouse wall in order to easily work on the connections and so that the circulation pump can fit.

Using treated lumber and plywood, shore up the walls of the trench between the sump and the baseboard of the greenhouse to create a service box for accessing the valves and the circulation pump. Make a lid out of ¾″ plywood. Alternatively, a prefabricated irrigation valve box can be used. If installing a cistern, make the service box large enough for both sets of pipes, with sufficient space for tightening four bulkhead fittings.

Connect the suction and discharge pipes to the inlet/outlet of the heat pump with threaded unions and gate valves on both. The suction pipe connects the sump to the inlet of the heat pump via the circulation pump. The discharge pipe connects the heat pump outlet to the sump.

Build a bypass port between the inlet and outlet pipes to adjust the water flow rate through the unit if necessary, and auxiliary ports between the gate valves and the unit for backflushing and cleaning the heat exchanger (see Chapter 11).

Heat pump and cistern valve service box.

Heat pump bypass, inlet filter and flushing ports.

Connect the 200 micron disc filter to the inlet port.

Run the suction/discharge pipes underground through the trench and under the baseboard of the greenhouse to the south wall of the sump. The height of the pipes is not critical, but they should run through the wall of the sump, not the collar. Once the connection location is determined, drill pilot holes through the sump, then use a hole saw from inside the sump. Install the clamp ring on the bulkhead outside the sump; install the gasket inside against the LDPE liner. Remember: socks only when standing or walking on LDPE.

Place ball or gate valves on both suction and discharge pipes in the service box outside the sump. Connect the circulation pump to the suction pipe using threaded unions. If the pump is powered by an outlet, run the power cord through an electrical conduit in the trench to the nearest outlet. If the pump is automatically switched by the heat pump, the HVAC tech will wire it to the heat pump control.

Connect the discharge pipe to the corresponding bulkhead and valve. Inside the sump, connect a short length of pipe and an elbow to the suction bulkhead in order to draw water from the bottom of the sump near the 4″ hydroponic return pipe. Install a swing check valve on the bottom of the suction pipe.

Cistern Installation

A cistern is optional but highly recommended. The cistern should be at least 4,000 liters (bigger is better) and located near the sump, outside the greenhouse, with the top at or below ground level of the greenhouse. This likely will require excavation. Use a 3–4″ layer of bedding sand under the cistern to ensure it is well seated.

A cistern serves three main functions:

1. It can be used to collect rainwater to feed the system.
2. It can be used as a storage tank for water from the system. This is a useful feature that can save a considerable quantity of water when doing a Salt Bath Treatment on the fish (see Chapter 8) or if you need to perform repairs.
3. It can be used as an overflow reservoir to prevent the sump from overflowing due to the Drain Down Effect in a power outage or pump failure (see Chapter 2).

A 2″ overflow standpipe in the sump, similar to the trough drains, will let excess water flow into the cistern when the water level in the sump is too high for any reason. A submersible pump in the cistern is used to pump water back to the sump as needed and is controlled by a switch or timer in the greenhouse. Rainwater can be collected in the cistern via gutters on the greenhouse which can connect into the overflow pipe or directly into the top of the cistern.

Once the cistern is in place, dig a trench for the 2″ overflow and 1½″ return pipes. Terminate the trench at the in-ground service box that was installed previously for the heat pump. Install the overflow and return pipes. The height of the return pipe is not critical, but the bulkhead for the overflow pipe must be at least 4″ below the intended maximum water level in the sump, with the pipe sloped gently downhill towards the cistern to drain properly. Run the power cable for the pump through a conduit in the trench and up the inside of the greenhouse on an arch. Later this will be wired to a switch.

Use a hole saw to install two bulkheads through the south wall of the sump, beside the heat pump connections. Drill from inside the sump with pilot holes first. Leave enough room between the bulkheads for the valves on each pipe to turn without contacting each other. Install the clamp ring on the bulkhead outside the sump. Install the gasket inside against the LDPE liner.

Connect ball or gate valves to the bulkheads on the outside of the sump, then connect the overflow and return pipes to the valves. Inside the sump, make a standpipe drain out of pipe and an elbow, like the trough drains, and attach it to the bulkhead fitting of the overflow pipe. The top of the standpipe should be slightly higher than the intended water level in the sump.

Connect the overflow pipe to the top of the cistern with a bulkhead or Uniseal. Drill a second hole in the cistern for the return pipe, just large enough for the pipe to slide through without allowing debris into the tank. Do not use a bulkhead or Uniseal in this hole as the discharge pipe needs to move freely up and down through the hole in order to install or remove the pump. The holes into the cistern must not be in the removable lid.

Once the plumbing layout is known, calculate the total head for the return pipe and source a pump with appropriate performance. You do not need a high-LPM pump but the slower the pump, the longer it will take to move water from the cistern. The pump must be submersible, with a float switch to prevent it from running dry.

Cistern installation. Note the service box between sump and greenhouse wall.

To install the pump, connect a riser to its discharge outlet so that it is just long enough to reach through the hole in the top of the cistern when the pump is sitting on the bottom. Install a swing check valve at the bottom of the riser to prevent water from the sump draining back into the cistern when the pump is turned off. Zip-tie the pump's power cable to the riser and secure

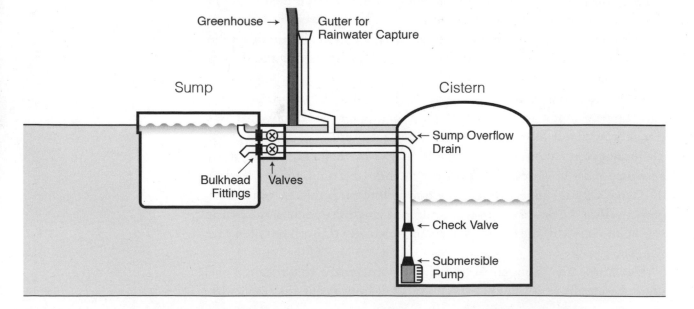

a length of waterproof rope (e.g., nylon) to the pump to assist with raising and lowering.

With the lid removed, lower the pump through the cistern's top access port, then slide the riser pipe and power cable back up through the predrilled hole so that the riser extends out of the hole by 4–6" with the pump sitting on the bottom of the cistern. Clamp a flexible rubber coupling to the top of the riser, then connect the riser to the return pipe in the trench with an elbow. If the pump ever needs to be serviced or replaced, simply unclamp the coupling, push the riser back through the hole and use the rope to pull the whole rig out of the cistern. Leave the rope hanging out of the access port and secure it so it cannot fall into the cistern.

Run the power cable through conduit into the greenhouse. Run the conduit and cable up one of the greenhouse arches. Wire a shutoff switch and a solid state timer in a weatherproof junction box attached to the arch. The timer will ensure that the pump cannot be accidentally left on, wasting electricity and potentially damaging the pump.

Seedling Table Construction

The Seedling Table will be located in the northwest corner. The table is 36" wide which allows for three trays to be fit on their 11" side (plus 3" of wiggle room). This will allow you to locate the table against the side of the greenhouse and still have easy access to all trays from one side. It is also an excellent width for supplemental lighting. Note that if using a tower system, a larger Seedling Table may be required due to the potentially larger quantity of plant sites.

The table can be constructed in any manner of your choosing, so long as it is sturdy. It should be one level, not stacked, and a comfortable working height, usually around waist high. A simple method, which we used, is to build a basic frame and legs out of 2×4 with ½" plywood as the table top. All edges of the table should have a 2" plywood lip. Fill all top seams and edges with a mold-resistant caulk. The table should be waterproof, which can be accomplished either with a membrane (LDPE works well) or by painting liquid rubber onto the table top and inside of the lips. We used liquid rubber. Build the tables so that they are slightly sloped towards one corner for easy drainage.

Place ¾" thru-hull drains in each of the outside corners where water pools and use ¾" hose to connect the drains under the table towards the wall of the greenhouse. Dig the hose into the ground under the greenhouse wall so that the water will percolate into the perimeter drain.

Install 400W HID lights per the layout shown in the DWC overview diagram on page 30. Each light should cover a maximum table span of 6′. The installation of the lights will depend on the manufacture of your chosen units. Two simple installation methods are to hang the lights on chains from the greenhouse frame or to build 2×4 supports over the table. Note that, unlike the supplemental lights over the troughs, these lights will not move laterally. The bulbs will be approximately 2–3′ over the plants. Ideally, you should be able to adjust the height as needed.

If using heating mats (which we recommend), install them now.

Workbench Construction

The workbench is a primary work area in the greenhouse. Seeding and transplanting are both done here. Construct a 3′×6′ workbench in any manner of your choosing, likely the same method as the Seedling Table: 2×4 framing with ½″ plywood top and 2″ plywood lip. This will comfortably fit a 2′×4′ raft with extra work space (needed particularly for transplanting). It should be a comfortable working height for you, likely the same height as the Seedling Table. Seal the bench with a membrane or liquid rubber. Slope the table slightly towards the north wall (no more than 1″ slope) and install a ¾″ drain via a thru-hull and hose to drain excess water into the perimeter drain.

For a tower system, we recommend a 3′×8′ table to allow for easy planting in 5′ towers.

Walk-in Cooler Installation

We recommend you consider a walk-in cooler mandatory. You will be producing large quantities of produce that are best stored in stackable totes in a cooler adjoining the outside of the greenhouse. If you are selling your fish fresh (not frozen), refrigeration is also required.

For the production you can expect from our design, we recommend your cooler be a minimum of 80 sq. ft. Coolers are very simple, and a good one should last many years if properly installed and not abused. The units consist of interlocking insulated wall panels that can be connected into different configurations. The refrigeration component will need to be installed by an HVAC professional.

New coolers can be extremely expensive. We suggest sourcing a used unit that is inspected by an HVAC professional prior to purchase. All modern units are air cooled, meaning they use air to cool the refrigeration pump. Some older units are water cooled and these tend to be inefficient, wasting huge volumes of water. Do not buy a water-cooled unit.

The cooler should be installed outside the west endwall. While it can be installed anywhere near the greenhouse, this location is the most convenient for loading. It can be located on either the north or south side of the west endwall doors. Ours is located on the south side; thus, we have shown it there in this book.

The cooler should be installed on a standard concrete pad (minimum 4"). When planning the location of the cooler, make sure you leave enough room between the cooler and the sliding greenhouse doors to allow room for the Germination Chamber.

We recommend that you hire a professional to build your cooler. The panels can be installed in a variety of layouts according to your needs, and the door can be located where you prefer. Install the door so that it can be accessed easily and without interfering with surrounding workspace. Our door is on the western end of the cooler.

IMPORTANT

The refrigeration system of the cooler must be installed by an HVAC professional. Do not attempt any work involving the refrigerant used in the cooler. The refrigerant can be toxic and is extremely dangerous.

The compressor on the roof of the cooler must be protected from the elements. The simplest method, which we used, is to install a basic lean-to roof off the endwall of the greenhouse. You can use any materials you deem appropriate. We used 6×6 posts with 2×4 rafters and metal roofing. The structure does not require walls; simply build an overhang of at least 2′ on all three sides.

Germination Chamber Construction

The Germination Chamber is attached to the side of the cooler nearest the greenhouse sliding doors, unless you plan on using a portable air conditioner to chill the chamber, in which case it can be built anywhere nearby. As discussed in Chapter 2, the chamber must be capable of holding at least 15 trays for our design. Additional trays may be required for a tower system. The Germination Chamber described here is capable of holding 18 trays on two levels (9 per level).

Both germination tables are lit with three 3′ T5 fluorescent lights (six total lights for the chamber). The lower lights are mounted to the bottom of the upper table; the upper lights are mounted to cross braces just below the roof. The lights are controlled by a timer. The temperature is controlled by a thermostatic controller.

The chamber is cooled by stealing cold air from the cooler. Cold air enters the chamber at the bottom, and warmer air returns to the cooler at the top of the chamber. If you opt to not install a cooler, you will need a small window-mount air conditioner to cool the chamber. The chamber is heated either via a small electric heater at the bottom or via heating mats under the trays.

No watering occurs in the chamber, so while no drains are needed, the levels should still be waterproof to prevent the wood from deteriorating. All electronics and controllers should be wall-mounted, and just like in the greenhouse, only GFCI outlets should be used.

The two 4″ holes cut through the cooler wall are fitted with backdraft dampers to ensure one-way operation and to minimize unintended air leakage between the chamber and the cooler.

Build the two 36″ by 72″ tables for the trays out of 2×4s and ½″ plywood using the same construction as the Seedling Table.

Build a frame out of 2×4s against the side of the cooler. The interior dimensions are 76″ wide by 44″ deep by 96″ high. The interior width/depth is greater than the table dimensions to allow room for insulation against the walls and air gaps at the front and back to allow air circulation.

Slope the roof slightly away from the cooler to shed rain and frame a set of doors on the front. The roof should overhang the three sides by at least 12″. The doors must be able to shut out drafts. Use weather stripping.

Sheath the walls and roof with ½″ plywood and caulk all the joints with silicone sealant. Shingle the roof and paint the sides and doors with at least three coats of exterior paint.

Construct the inner bracing out of 2×4s to hold the tables. Both tables need a 2–3″ gap at the front (between the tables and doors) and back (between the tables and cooler) to allow air circulation. The lower table is 30″ off the ground if using a space heater. It can be lower if seedling mats are used. The bottom of the upper table frame is 24″ from the top of the lower table. This is the appropriate distance to mount the lower lights. Install two cross braces across the chamber (left to right) 24″ above the top of the upper table to mount the upper lights.

If you are using a window-mount A/C unit to cool the chamber, install it in a side wall above the upper table. Use small fans to circulate the cold air around the chamber.

Before installing the tables, caulk all interior seams and joints, then insulate the sides, doors and roof by gluing 2″ styrofoam panels to the inside.

Be precise with the cuts to avoid gaps. No insulation is required against the cooler.

Run power to the chamber. Use GFCI outlets. The chamber and the cooler evaporator fan (inside the cooler) can share a circuit if the fan is 120v. Typically a cooler compressor (on top of the cooler), which is a large electrical load, is 240v and requires a dedicated circuit.

Use a hole saw to cut a 4″ hole through the cooler in the center bottom of the back wall, below the lower table. Cut another 4″ hole in the top center of the chamber, above the cross braces for the upper lights. Install a 4″ backdraft damper in the top hole inside the chamber. This allows air to only flow into the cooler. Install a second damper in the lower hole inside the cooler. This allows air to only flow into the chamber. Mount a 4″ in line fan (minimum 100 CFM) onto the bottom damper so that it blows cool air into the chamber. The air is forced up around the tables and out the upper hole back into the cooler.

Install the tables leaving gaps at the front and back for circulation.

Mount the thermostat against the back wall, just below the bottom of the upper table, and connect the fan and heater to it. Plug the thermostat into a GFCI outlet.

The fluorescent lights can either be chained together for power or plugged individually into a power bar. In either case, the lights are connected via a timer plugged into a GFCI outlet.

Tools of the Trade

5

By THE END OF CHAPTER 4, you should have a completed greenhouse with all major construction completed. The greenhouse is fully built and installed, including troughs, tanks, filter and sterilization systems, seedling and work areas, walk-in cooler with adjoining Germination Chamber, and all plumbing, aeration and electrical. Fully oxygenated water, filtered by the filtration units and sterilized by the UV units, should flow freely through the whole system when the pumps are turned on (see Chapter 7).

This chapter is a component list for the additional major tools that will be used in the system. With the exception of the freezer, we consider all tools in this section to be mandatory for consistent long-term success.

Shade Cloth

Shade cloth is a woven netting that reduces the intensity of light during the hotter times of year. It is typically made of polyethylene (PE) and comes in various opacities that specify the amount of light blocked by the cloth (30% cloth will block 30% of the light that hits it). Good-quality shade cloth is customizable in its dimensions, sewn on all edges, UV stabilized, has numerous grommets to serve as anchor points and has a life expectancy of at least 10 years.

The use of the shade cloth will be highly dependent on your latitude and local environmental conditions. In general, we recommend installing shade cloth when daily temperatures are above 25°C (77°F). On our farm in southwestern Canada, we usually have cloth installed from around early July to early September.

In temperate or colder climates, 30% opacity will usually suffice. In warmer climates where the sun is more intense and for longer daily photoperiods,

Our 50% opacity shade cloth.

we recommend 50% opacity. The shade cloth should be sized to cover your entire greenhouse, down to ground level, excluding the endwalls (unless your greenhouse is oriented north-south, in which case the south endwall requires shading). It will be one large cloth and will anchor along the long sides of the structure.

Our cloth is made by American Nettings & Fabric, Inc. There are numerous other companies that produce similar products.

Install the shade cloth by laying it out along one side of the greenhouse, affixing ropes to several grommets and throwing the ropes over the greenhouse. Pull the ropes from the opposite side to situate the cloth, taking care not to snag or tear the cloth or greenhouse covering. At the end of the hot season, uninstall the cloth when it is dry and store in a dry location.

Backup Oxygen and Power

One of the greatest threats to your farm is a power outage. An aquaponic farm relies on constant electricity for operation. Periodic power outages happen in all power grids, so preparing for them is essential.

The single most threatening consequence of a power outage is oxygen depletion. Your system can survive for a prolonged period without water movement and filtration, but the oxygen in the water will be rapidly consumed by the fish, plants and bacteria. Fish in particular require constant high levels of oxygen and will likely start dying within 30 minutes without replacement oxygen. This holds true for all types of systems including towers.

One solution is to install a backup generator system. There are numerous types of generators available for such purposes, and you would need to size yours according to your electrical load. Not all items need to be connected to the generator, as not all are vital. The most crucial items to back up are the aerators, water pumps, propane heater and, in the case of double-poly greenhouses, the poly inflation blower.

The downsides to a generator are that they can be very expensive and, like all internal combustion engines, are not 100% reliable. Rather than using a generator, we recommend installing a backup oxygen system. Even if you opt to install a generator, we recommend also installing a backup oxygen system.

In addition to being the best remedy for power outages, a backup oxygen system is also very handy for supplementing oxygen to fish tanks during times when the tanks are taken offline from the main system, such as when doing a Salt Bath Treatment or when new cohorts are quarantined.

A backup oxygen system is very inexpensive and is virtually 100% effective in a power outage. The backup system is simply compressed tanks of pure oxygen that are set to provide oxygen to the fish tanks and the MBBR if the power goes out. The MBBR is vital as it is the largest host for bacteria which also consume considerable quantities of oxygen. The frequency and length of power outages in your area will dictate the volume of the oxygen tanks. We suggest having at least two "T" or "K" size tanks so that you can swap them out for refilling as needed. One "T" or "K" size tank holds enough oxygen to keep the fish and bacteria alive for about 24 hours when using micro-bubble diffusers. We rent our tanks from a local welding shop and swap them out for fresh tanks as needed.

The tanks must be located outside the greenhouse in a well-ventilated area that is protected from the weather. The tanks must not be located in the greenhouse or anywhere near fuel-burning appliances.

The tanks are connected to a pressure regulator that controls the outlet pressure. Set the regulator at 40–50 psi. The regulator is connected to a solenoid valve that is "normally open" (a valve that is open when the solenoid is not powered). During regular operation, with power running to the solenoid, the valve is closed and no oxygen flows. When power no longer reaches the solenoid, as in a power outage, the valve automatically opens, allowing oxygen to flow.

The solenoid valve should be wired by your electrician to a switch and power supply. We use and recommend Jefferson solenoid valve # ZC-1327BN402TINA-O. If using any other solenoid valve, make sure it is the correct voltage and operating pressure and that it is specifically rated for use with oxygen.

The outlet of the solenoid valve is connected to a five-port flow meter manifold which can be located inside the greenhouse. One port controls the oxygen flow to the MBBR, and the other four control flow to each of the four fish tanks via ¼″ vinyl tubing. Run the tubing through conduit from the flow meter manifold, overhead through the arches, dropping down into each tank and the CFB. Drill a ¼″ hole at the top center of the east side of the CFB for the tubing. Once in place, seal the inside and outside around the tube with epoxy resin.

Backup oxygen system: oxygen tank, pressure regulator, solenoid valve and flow meter manifold.

To increase the efficiency of oxygen transfer and extend the operating time of the backup oxygen system, use ultra-fine pore diffusers rather than the medium-pore diffusers used in the troughs. We recommend Sweetwater Micro Bubble Diffusers.

In the event of a power outage, it is still crucial that you immediately go to the greenhouse to confirm the backup oxygen is working. The backup system recommended here is very simple with little chance of failing, but the consequences are dire, so you should always double-check. You will know that the power is out because of the monitoring system.

We recommend linking the backup oxygen system with the monitoring system to automatically trigger the oxygen in events other than a power failure, such as pump or aerator failure.

Monitoring System

A monitoring system is essential for your farm. One of the primary benefits to aquaponic production is how little work is required to produce large quantities of food compared with traditional soil farming. The flip side of this is that the majority of the 168 hours in a week are not spent in the greenhouse. Having a monitoring system to immediately alert you to something going wrong is therefore essential regardless of the backup systems you have in place. You need to know if something is amiss.

There are numerous monitoring systems on the market that you can use. Essentially what you require is an autodialer that you program with specific parameters to notify you if it reads anything outside those parameters. Some are connected to the internet and use data to contact you, some use telephone or cellular networks, and some connect via satellite. Internet connections will be the cheapest to operate; satellite the most expensive. There are a variety of options for how to be notified, including text message, email or phone.

We use and recommend the Web600 monitoring system by Sensaphone. It is a small network-enabled unit that connects to a router and can send text message or email alerts to any device via the internet. Internet service onsite is required. The Web600 has a row of contact terminals to which a variety of switches and probes can be connected, such as flow or pressure switches, temperature or humidity probes, and smoke alarms or motion detectors.

It also has a built in SPST relay output which is capable of triggering the backup oxygen system. This is particularly useful in situations other than a power failure that would result in low oxygen conditions, such as pump or aerator failure. The Web600 is programmed by accessing the device from a web browser on your computer and can be set up to escalate alarm notifications to multiple contact profiles.

The crucial parameters to monitor are power, water flow and aeration. Optionally, you can also monitor water temperature, air temperature, motion (for security) or many other things of your choosing. Power, water flow and aeration are mandatory.

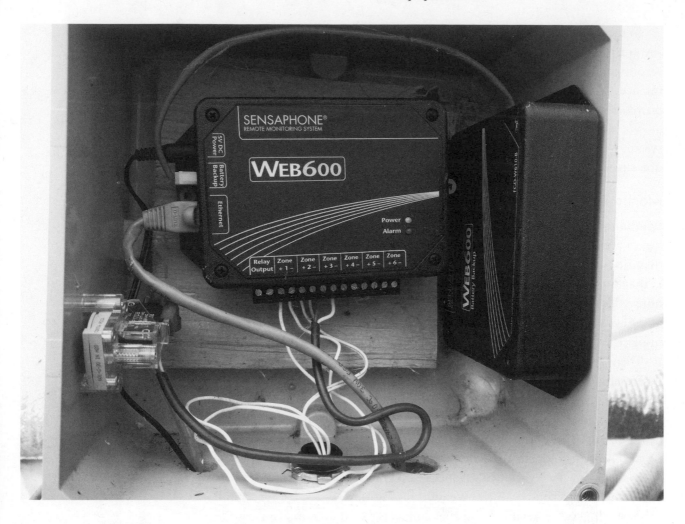

Web600 monitoring system with battery backup (right) and aeration pressure switch (left).

To monitor power, you will need a battery backup (this is an extra item with the Web600). Note that the network router the monitoring system is connected to must also have a battery backup (UPS) for alarms to be sent in a power outage.

Water flow is monitored with a flow switch located between the pumps and the UV inlet manifold. We recommend the Aqualarm 10235 Flow Detector which is a simple on/off paddle switch designed to be drilled and tapped into a pipe or threaded into the branch of a tee fitting. If the flow drops below 40 LPM, the switch opens and an alarm is triggered.

Aeration is monitored with an air pressure switch connected to the air distribution system. Connect an extra length of ¼″ tubing to the air distribution pipe located in the sump. Run the tubing through the west end of the sump near the monitoring equipment and connect it to the high side of the pressure switch. We recommend using a low-pressure diaphragm switch such as the MDS-6 (compact pre-set) or the Model 1638-10 (adjustable) from Dwyer Instruments. The pressure switch must have a setpoint of 10″ WC (Water Column) or less. Air pressure in the distribution pipe keeps the switch in an open position. If the aerator fails for any reason, the drop in pressure will close the switch and the monitoring system will send an alarm notification.

We highly recommend linking the monitoring system with the backup oxygen system in case of pump or aerator failure. The built-in relay output in the monitoring system can be programmed to trigger electronic equipment such as the oxygen solenoid valve. To do this, your electrician will need to wire a "normally closed" relay switch (with a control voltage of 30v/1A or less) into the backup oxygen power supply and run a control wire to the Web600. During normal operations, the relay will allow power to flow to the solenoid valve. When a programmed alarm occurs, the monitoring system will send a control signal to the relay which will open the switch and cut off power to the solenoid, which in turn will open its valve and send oxygen to the tanks.

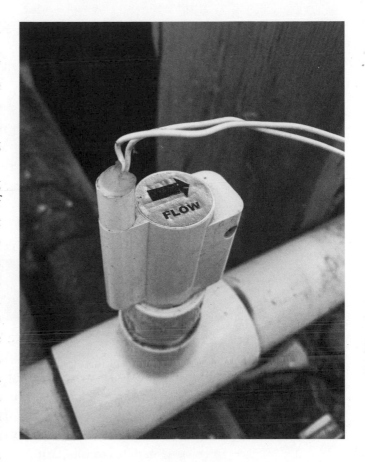

Aqualarm 10235 Flow switch.

pH Controller

pH monitoring and control is critical for the system to thrive. It is not acceptable to just make daily manual corrections. There are only a few commercially available monitors that monitor and control pH. We use and recommend the Bluelab pH Controller which has an integrated peristaltic pump with user-replaceable components, and when calibrated properly, we have found it to be very reliable.

Calibration is simple and done monthly. Use only Bluelab calibration standards. We have found other brands of calibration standards, although cheaper, to be inaccurate. In addition to regular calibration, it is mandatory that you use chemical pH tests to verify that the electronic monitor is reading correctly as, even if regularly calibrated, it is possible for electronic monitors to give inaccurate readings. Chemical kits, while less precise than electronic monitors (as the reading is given in a color gradient), are far more reliable to confirm the approximate pH.

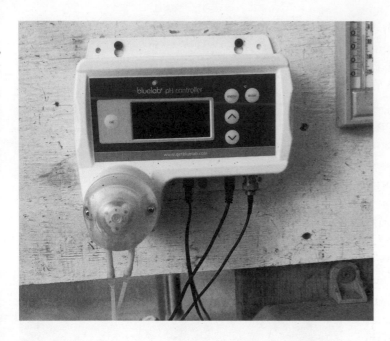

Bluelab pH Controller.

pH is monitored in the sump with the probe located near the pump intakes (just before the water leaves the sump). The pH adjusting chemicals should be introduced to the sump at the east end, near the return from the hydroponic troughs, to allow proper mixing with the system water. The tubing that is supplied with the controller to adjust pH is run along the inside of the sump collar to the east end of the sump near the return from the hydroponic troughs for distribution. Keep the end of the tube above the water.

Dissolved Oxygen Meter

A Dissolved Oxygen (DO) meter monitors the level of available oxygen in water. It is used to measure the oxygen in the fish tanks (including the Purge Tank) and when transporting cohorts of fingerlings.

When we first started, we believed this to be a crucial piece of hardware and bought a very expensive unit. We do not recommend this unless you plan to keep an ongoing log of your DO levels (we do not). While it is important to have a DO meter at the establishment of the system and to transport fingerlings to your farm, you will not require an expensive unit capable of a very high degree of accuracy.

We use an Orion Star A2235 portable DO meter that cost more than C$1,500. We must calibrate the unit before every use, and the very expensive probes must be replaced annually. We do not recommend it or units like it. An example of a meter we recommend is the Oxyguard Polaris DO meter, which costs approximately C$900.

Water Testing Kit

Water testing kits are mandatory and need not be complicated or fancy. Your kits must be able to test for ammonia, nitrites, nitrates, iron, phosphorous, calcium, hardness and potassium. It is uncommon for potassium testing to be including in a general testing kit, so you will need a stand-alone potassium kit.

We recommend the Nutrafin A7860 master test kit as it contains everything you need except the potassium test. We also recommend the API Freshwater Master Test Kit. The API kit tests only for pH, ammonia, nitrites and nitrates, but we have found the pH and ammonia tests to be more accurate than the

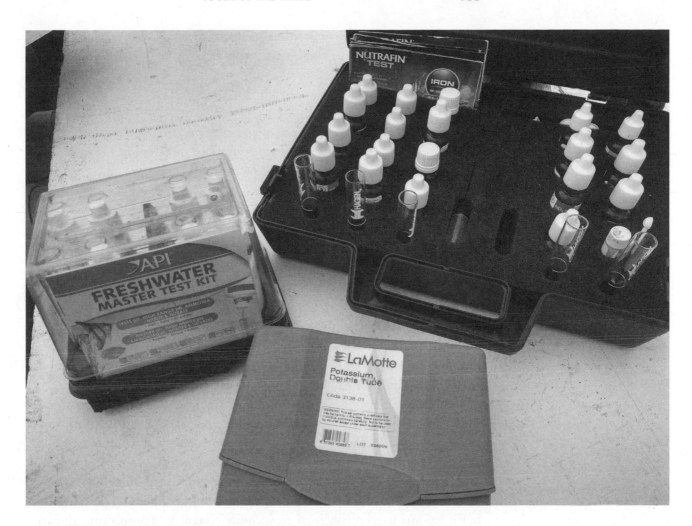

Nutrafin. For potassium, we recommend the Lamotte 3138-01 Potassium Kit.

Our water testing kits.

In general, testing kits should be replaced one year after opening. Like many of the suggested components, we recommend purchasing from Pentair Aquatic Eco-Systems.

Seedling Trays and Domes

You will be using standard 11″ × 22″ nursery trays for germination and seedling growth. Our design is based around trays that contain 98 cells per tray. We find these to have the ideal spacing for growing seedlings up to transplant size without overcrowding or wasting space.

There is a variety of quality in trays. Cheaper trays will wear out more rapidly and may be more frustrating to use. You will need to experiment to discern which trays work best for you. Trays are often available at a wholesale rate for cases of 500. No matter the quality of the tray or how you use them, they will wear out, so we recommend buying a case once you have determined your desired tray.

Substrates

You will need two different types of substrate: a growing substrate for germination/seedlings and a substrate for transplanting seedlings into net pots. Substrates must be free of residual salts (common in cheaper coco coir mixes) and pH neutral.

The common choices for a growing substrate are a peat-based potting soil such as Pro-Mix or coco coir (ground up coconut husks). We have found both to work. We use Pro-Mix for the growing substrate as we have found it provides better germination and it is less expensive. The downside is that peat-based mixes are less environmentally friendly as, unlike coco coir, they are non-renewable.

The transplanting substrate is used when transplanting seedlings into net pots. It helps to prevent the growing substrate in the seedling plug from eroding into the system due to the vigorous mixing action of the air stones in the troughs. Erosion causes poor growth and an accumulation of growing substrate on the bottom of the troughs which leads to anaerobic zones and impaired nitrification as the biofilm is smothered. The transplanting substrate, which is placed in the bottom of the net pot before inserting the seedling, must be coarse enough to withstand the eroding action of the air stones.

We have experimented with a variety of transplanting substrates. Mineral wool, also called rockwool, is the best at holding the seedling plug together, but it has an unacceptably high environmental cost to manufacture, does not decompose in compost and cannot be used in organic production. Expanded clay pebbles work well and are reusable but require additional labor to recover them after harvesting and to wash them before reuse.

Our choice for transplanting substrate is very coarse coco coir chips which keep the seedling plugs intact, can be composted with the root balls after harvest and are an environmentally friendly renewable resource.

Dibbler Plate

To speed up the process of "dibbling" (poking shallow holes for seeds in growing media), make a dibbler plate that can do a whole tray with one push.

1. Cut two 11″×22″ pieces of ½″ plywood.
2. Use a seedling tray to mark the center of each cell through the drainage holes onto one piece of plywood.

3. Drill ¼″ holes through each mark.

4. Glue the second piece of plywood to the first. Ensure they are aligned and clamp them while drying.

5. Before the glue sets, affix a 1″ long piece of ¼″ dowel rod in every hole with a dab of glue on the end of each.

6. Once all glue dries, saturate the dibbler plate with epoxy resin to waterproof it and allow to cure for 24 hours before use.

Net Pots

Our design uses 2″ net pots, sometimes called slit pots, which are 2″ wide at the top by 2″ deep and have a small lip around top edge to hold the pot in the raft. You will need enough for the system plus at least 25% more as some will always be in the cleaning process. For our design, you will need at least 7,500 net pots.

Totes

Totes are a common item on most family-sized farms. They are used for a wide variety of purposes from harvesting and storing produce to transporting items to customers. The size of the totes depends mostly on what is comfortable for you to move when loaded. In general, we recommend totes that are between 40–70 liters. We suggest purchasing higher-quality totes made of non-rigid plastic. Rigid plastic will crack and split with ongoing UV and cold exposure. We recommend Rubbermaid Roughneck Storage Boxes which are highly durable and typically retail for under $10.

Packaging

We tried many things to avoid using plastic bags, but in our experience they are a necessary evil. Plastic bags will keep delicate herbs and salad mix from drying out during transport, and most end customers will want a bag when selling at a market. Unless you are selling in bulk to a large wholesale buyer, you will likely require plastic bags. What you require will depend on your market and/or how you are marketing your food. We use several sizes of bags: small for preweighed items such as herbs, mid-size for salad mixes and larger for head lettuce, romaine hearts and living butter lettuce.

Dibbler plate.

Our well-used salad spinner.

There are plant-based alternatives to plastic that are acceptable for storing food, but they tend to be expensive, weak and not always as environmentally friendly as they might seem.

Salad Dryer

A commercial size salad spinner/dryer is necessary if you plan on making salad mixes. We use a DynamicTM 20 liter manual salad dryer which is also available with a motor. Salad dryers are best sourced from restaurant equipment suppliers.

Scales

You will use two scales: a low-weight higher-resolution scale (~20 kg ×1 g) for weighing produce and individual fish and a high-weight lower-resolution scale (~100 kg × 20 g) for fish sampling (see Chapter 8). We recommend splash-proof or waterproof bench scales. Analogue (sliding beam) and digital scales are both fine but if not resistant to moisture will rust and malfunction eventually in the humid greenhouse environment.

Knives and Scissors

Knives and/or scissors will be used frequently, especially for harvesting. There are a wide variety of options available. We have tried many types of knife and prefer using high-quality harvesting scissors.

Fishing Nets and Tank Covers

You will need four fishing nets for grading, transfers and mortality removals. One net will be dedicated to each tank to prevent the spread of disease from tank to tank. Look for a long-handled monorail type net, with a deep, knotless bag, such as the DN33D Monorail Net from Pentair.

Tank covers are recommended as they keep frisky fish from leaping out of the tanks (which can happen surprisingly often), and they provide extra shade and comfort for the fish which helps to prevent stress. If you are handy with a sewing machine, they can be stitched together from shade cloth and a drawstring, or they can be purchased from Pentair.

Cleaning Brushes

You will need the following brushes:

- 4 brushes with long handles, one dedicated to each fish tank. The brushes must be able to reach the drain of the tank; 6′ handles are just about right.
- 1 brush with a long handle for cleaning the inside and bottom of the RFS. A 1–2′ handle is ideal.
- 1 or more short-handled brushes for general maintenance and cleaning.
- 1 or more long-handled brushes for cleaning rafts (depending on how many people do this job at the same time).
- 1 regular dish brush for cleaning harvested fish.

All of the brushes should be soft bristled. Do not use metal or rigid plastic bristles for any job.

Washing Machine

A washing machine is optional but highly recommended to wash net pots. It is possible to do this job by hand, but it will add considerable labor. Only use unscented, non-toxic, biodegradable laundry soap.

One of our homemade fish tank covers.

A standard top-load machine works very well. You should be able to find one used for very little money. The washing machine should be located outside of the greenhouse but as near to the doors as possible. Ours is located next to the Germination Chamber.

Chest Freezer

A chest freezer is optional but very useful. We use ours primarily to store frozen fish. It can be located wherever is appropriate for you. Ours is located on the opposite side of the cooler from the Germination Chamber and washing machine.

Hot Water System

Hot water in the greenhouse is optional but may be required for a food and safety permit. The two options are a traditional hot water tank or an on-demand system. On-demand typically operates on propane — we recommend it rather than a tank.

Stainless Steel Counter/Sink

For a variety of tasks, from washing hands and plants to cleaning fish, you will need a work area that is designed to get wet and is easy to keep clean and sterilize. An all-stainless counter with a built-in sink is the best option and the

Our counter and sink.

only one we recommend. We purchased ours used for a fraction of the price of a new unit.

The sink will need to drain per your local building code. The sink cannot drain into the Waste Tanks as you will be using sanitizers, bleach, soap and water quality testing chemicals.

Constructing the Wash Down Sink

The wash down sink is where you will wash the CFB filters screens 2 or 3 times per week. To capture the solids for fermentation, build a simple vessel of your own design that drains into the primary Waste Tank.

Feed Storage

Commercial fish feed typically comes in 20 kg bags, and storage must be dry, vermin-proof and as cool as possible. The storage does not need to be climate controlled unless you are buying more than a 12-month supply, which we do not recommend, and it does not need to be fancy. We built our feed storage container out of scrap lumber and plywood on a pallet, insulated the inside with sheets of styrofoam and waterproofed it with a few coats of white exterior paint.

Our wash down sink, constructed of plywood and coated with liquid rubber, plumbed into the Waste Tank.

Our feed storage container.

Managing the Ecosystem

6

CHAPTERS 1–5 INTRODUCE all the major components and concepts of an aquaponic farm and teach you how to build our design. Chapters 6–11 will discuss the major subsystems (aquaculture and hydroponics) and teach you how to run the farm.

In this chapter, we discuss the primary job of an aquaponic farmer: managing the ecosystem you have created. In particular, we will discuss how bacteria are the vital link between fish and plants.

Bacteria

Most people understand that aquaponic farming means raising fish and growing plants. While this is true, we encourage you to think of yourself first as a bacteria farmer. The much-vaunted relationship between plant and fish is actually a three-way symbiosis with bacteria as the vital middle link. Without vigorous bacterial colonies, no amount of work will permit the fish and plants to survive, let alone thrive.

The role of bacteria in an aquaponic system is to remove or decompose the metabolic wastes produced by raising fish. This is called biofiltration. Wastes include ammonia, which is excreted from fish gills and is also found in fish urine, fish feces and any uneaten fish feed. Removal and decomposition of wastes from the system is critical. Ammonia quickly becomes toxic to fish at levels greater than 3 ppm, and high levels of organic particulates can also be detrimental to fish health.

Oxidation of ammonia by bacteria is called nitrification. The decomposition of organic particulates into their elemental constituents by bacteria is called mineralization. The bacteria colonies of an aquaponic system are called the biofilter.

Nitrifying Bacteria

The job of nitrifying bacteria is to convert ammonia into nitrates through a two-step oxidation process accomplished by two families of nitrifying, bacteria: ammonia oxidizing bacteria which oxidize ammonia into nitrites, and nitrite oxidizing bacteria which oxidize nitrites into nitrates. There are many species and strains of bacteria within each family. In an aquaponic ecosystem, the primary ammonia oxidizing bacteria are those of the genus *Nitrosomonas,* and the primary nitrite oxidizing bacteria are those of the genus *Nitrobacter.*

Nitrates are a bioavailable form of nitrogen which is the most important mineral element for the growth of plants, particularly in the leafy vegetables you will be growing. When the biofilter is robust and operational, the two stages happen almost simultaneously, which is to say that the nitrites, once created, are very rapidly converted into nitrates. A high level of ammonia (>2 ppm) and/or nitrites (>1 ppm) is an indication of a problem with the biofilter.

Nitrifying bacteria are autotrophic: they derive their energy from inorganic compounds, primarily ammonia and CO_2 in an aquaponic system. The bacteria involved in this process build static colonies called biofilms that grow on all the underwater surfaces. Biofilms are slimy films that coat all dark and well-aerated underwater surfaces. They are obligate aerobes (they require oxygen to grow) and are relatively non-mobile, tending to colonize an area and stay there. Nitrifying bacteria are slower to reproduce than mineralizing bacteria and can be quickly out-competed for space and oxygen in the presence of high levels of organic particulates.

Mineralizing Bacteria

The job of mineralizing bacteria is to convert the suspended organic particles in the system into other essential minerals vital for plant health. Mineralizing bacteria are heterotrophic: they derive their energy from organic compounds, primarily feces and excess feed in an aquaponic system. They are a large family of thousands of different species. Unlike nitrifying bacteria which create immobile colonies, mineralizing bacteria float freely throughout the system, colonizing and decomposing suspended solids. Mineralizing bacteria reproduce much more rapidly than nitrifying bacteria. As mineralizing bacteria can be aerobic or anaerobic, the mineralization process occurs in all areas of the system where solids accumulate, even in areas of low oxygen, such as the RFS, filter screens and tank drains.

Bacteria and UV

Both classes of bacteria are killed by the UV levels we recommend. For nitrifying bacteria which are mostly static, this is not a problem as only outliers to the colonies pass through the UV units. For mineralizing bacteria which reproduce rapidly, the UV sterilizers help to keep the population at an appropriate level.

Mineral and Nutrient Content

Ammonia, Nitrites and Nitrates

Ammonia, nitrites and nitrates are all forms of nitrogen. Ammonia is toxic to fish at short-term concentrations of >3 ppm and long-term concentrations of >1 ppm. Nitrites are highly toxic to fish at concentrations of >1 ppm. Nitrates are far less toxic to fish and can be tolerated at concentrations up to ~1,000 ppm, although under normal conditions, the plants will keep nitrates below 500 ppm.

Plants can tolerate higher concentrations of ammonia and nitrites than fish and can tolerate very high concentrations of nitrates (> 2,500 ppm), though beyond a certain point they are redundant. Nitrates are the primary source of nitrogen for plant growth, so the plants will suffer from nutrient deficiencies when nitrate concentrations are lower than 100 ppm for prolonged periods. The sensitivity of the fish to nitrates is the limiting factor when it comes to nitrate concentrations.

IMPORTANT

The key tolerances for our system are: ammonia less than 3 ppm, nitrites less than 1 ppm and nitrates between 100–1,000 ppm.

Other Plant Nutrients

The crucial elements for plant growth are nitrogen (N), phosphorous (P), potassium (K), calcium (Ca), magnesium (Mg) and iron (Fe). N, P and K are typically considered macronutrients and all others micronutrients. In an aquaponic system that grows only leafy (non-fruiting) plants and that must sustain bacterial colonies, the vital nutrients are N, K, Ca, Mg and Fe. Phosphorous, though present in small amounts, is not needed in the large quantities required for fruiting plants.

The primary input into the system is fish feed. It is the origin of all of the nitrates and the majority of minerals. Fish consume the feed, excrete ammonia through gills and urine, produce large amounts of feces and leave some feed uneaten. The ammonia is converted into nitrates via a two-step oxidation nitrification process which is the primary source of nitrogen for the plants. The feces and excess feed, converted into their elemental constituents via mineralization, are the source of most other minerals and nutrients.

Although the feed contains everything needed by the fish, it tends to be deficient in Ca, K, Fe and occasionally Mg. For ideal plant and bacteria growth, these elements must be supplemented to keep the mineral content balanced.

Ca and K are supplemented when added to the system in hydroxide form to manage pH (see below). Hydroxides are very alkaline, thus adding them in this manner has the double benefit of raising the pH and supplementing deficient minerals.

Fe and Mg don't affect pH and tend to be less deficient in most systems, depending on the water source. They are supplemented less frequently by adding a small amount (approximately 1 ppm) of Epsom salts (magnesium sulfate) and 2 ppm of chelated iron about every 2–3 weeks.

IMPORTANT

Only supplement with a high-quality chelated iron. Never input raw iron or any iron that can rust. Iron oxide (rust) is toxic to all organisms in the system.

The mineral content in your source water will affect the balance of minerals in the system and can vary greatly depending on the source. Regularly test your water source and adjust accordingly.

The balance and concentrations of all minerals and nutrients are monitored through regular testing with a chemical testing kit and regular observations of the plants.

Chemical Testing

Chemical tests use a chemical reaction (titration) to measure the concentration of specific elements in a water sample. They do not rely on electronics, and the only potential for incorrect measurement is human error. We use chemical tests on a weekly basis to assess ammonia, nitrites, nitrates, potassium, calcium, iron and, optionally, phosphorus. During the initial cycling, or startup phase, you will be testing often (up to twice per day). As your system stabilizes and you become familiar with its tendencies, testing frequency decreases to once or twice a week.

As discussed in Chapter 5, testing is done with a combination of three kits. We use an API Freshwater Master Test Kit to test ammonia, nitrites, nitrates and pH, and to confirm the accuracy of the pH controller. We use a Nutrafin A7860 Master Test Kit to test for hardness, calcium, phosphorus and iron. We test for potassium with a Lamotte 3138-01 Potassium Kit.

Test pH weekly with a chemical test to confirm that the digital pH controller is accurate. All digital controllers require regular calibration and even when regularly calibrated can read incorrectly. Chemical pH tests are more reliable than digital meters as long as the tests are not past their shelf life and the operator carries out the instructions correctly. When calibrating a pH controller, use high-quality calibration solutions. We recommend Bluelab 4.0 and 7.0.

The procedure for using a chemical test kit varies from kit to kit. Follow the instructions precisely.

Sample Water Location

The location from which sample water is taken for testing is important as different areas of the system will show different results. For example, if testing ammonia for biofilter performance, collect the sample from one of the inlet spouts (immediately before the fish tanks). The primary source of ammonia is the fish, so this location will have the lowest ammonia, as the biofilter has the most time to oxidize it. Ammonia should not be present in high concentrations at an inlet spout if the biofilter is functioning properly. Conversely, if a test sample is taken from the Tank Manifold (immediately after the fish tanks), the test will show elevated levels of ammonia which may incorrectly be interpreted as a sign of poor biofilter performance, when in fact the sample water hasn't yet been treated by the biofilter.

IMPORTANT

All samples for fish water quality testing are collected from the tank inlet spouts. All samples for plant water quality testing are collected from the trough inlets.

It is also possible to estimate the efficacy of each component in the system by collecting samples at both the inlet and outlet of a component and calculating the difference between them. For example, to determine the concentration of ammonia the fish are producing at a given time, collect water samples at a tank inlet spout and at the tank's SPA outlet. If the inlet sample tests at 0.25 ppm and the outlet tests at 2 ppm, we can estimate that the fish are producing around 1.75 ppm of ammonia at the time the test was taken. Using this same method, it is possible to estimate the nitrification rate of the biofilter, the denitrification rate of the hydroponic subsystem and the oxygen consumption in each fish tank.

Plant Observation

In all types of farming, being proficient at gauging plant health via direct observation is crucial. Over time, you will become an expert at identifying bug and pathogen issues, mold and mildew, and nutrient deficiencies before they become a serious problem. Daily observation and early removal or treatment is the number one tool in the greenhouse and the first line of defense against all problems.

Every day when you arrive at the greenhouse, walk through the troughs and observe your plants. This should only take a few minutes. On a weekly basis, do a more in-depth plant evaluation where you look at each plant for problem symptoms.

Does anything look different from yesterday? Do you see any problem symptoms, such as odd discoloration in the leaves, unusual or unexpected growth patterns, pests or signs of pests, or any mold or mildew?

We cover this subject and remedies in Chapter 10. The key point is that certain signs, particularly leaf discoloration, are indicators of a lack of one or more nutrients. Regardless of what the chemical tests tell you, the plants are the true arbiters of the mineral quality of the water. If the plants are healthy and robust and growing rapidly, strive to maintain the mineral content as it is. If the plants show deficiencies, discern what is missing and promptly supplement.

Nutrient Recycling

If you opt to produce your own biodigested fertilizer, one benefit is that it can be added back into the system to supplement minerals and boost the overall nutrient content. The fertilizer created in the biodigestion process contains numerous minerals that either weren't previously bioavailable to the plants or were only present in lower concentrations. By adding your own fertilizer back into the very system that created the starting material, you can "close the loop" by recycling what was previously a waste material and, in the process, create an even more sustainable aquaponic farm.

The amounts to recycle will vary greatly, based on the concentration of your fertilizer and its primary constituents. Lab test the fertilizer to evaluate how much is appropriate. The testing is also very helpful, perhaps even mandatory, for selling it. As a rough reference, we recycle approximately 8 liters every week back into our system. The only limiting factor to how much can be recycled is the requirements for fish water quality. As long as there is no ammonia present in the fertilizer and it is well aerated, you should have no problems. When recycling fertilizer back into the system, divide it evenly into the trough inlets.

Water Quality and Management

Fish, plants and bacteria (nitrifying and mineralizing) all have different water quality preferences. The objective is to strike a balance between the requirements of each while maintaining an environment that can be tolerated by the individual organisms. As the goal of a commercial aquaponic farm is to generate revenue from plants, and to a lesser degree fish, the balance between the conditional requirements is heavily weighted in favor of the elements that actually generate revenue. The essential qualities to manage are pH, water temperature, dissolved oxygen, nitrites and nitrates, and mineral content.

pH

Source water will vary in pH. In general, municipal source water will be between 6.5 and 7.5. Well water will vary dramatically depending on the local geology and may fluctuate seasonally. Water under 6.0 is extremely acidic and over 8.0 is extremely alkaline. Avoid using source water that is either extreme.

Due to the continuous bacterial processes of nitrification and mineralization, the pH of an aquaponic system will typically slowly acidify. Without correction, the system will eventually become too acidic for fish and bacteria. Plants will also suffer from nutrient deficiencies if the pH is too low or too high for an extended period.

Each organism prefers a different water pH. Fish in general prefer a pH of 7.0 or higher, though certain species, including trout, will tolerate and still thrive in slightly acidic water. Plants on the other hand require acidic water to fully utilize the wide variety of minerals needed for growth. Both types of bacteria also prefer a pH of 7.0 or higher, and as the pH drops, their growth and performance increasingly suffer.

When considering the different pH preferences, we prioritize the needs of the plants as long as the fish can still thrive and the bacteria can perform well. Depending on the quality of feed used and the price per pound for harvested fish, you will more or less break even on raising fish. Plants generate the vast majority of aquaponic income, and the more robust and rapid their growth, the more income.

IMPORTANT

Our system is designed to operate at a pH of 6.5, +/- 0.1. This pH is ideal for leafy green plants, is well tolerated by trout, and while outside the optimum range for both types of bacteria, they still grow and perform more than well enough to serve their functions.

Two pH Management Methods: Buffering vs Hydroxides

There are two primary methods of pH management in an aquaponic system: "buffering" the pH or "adjusting" the pH.

"Buffering" refers to water's ability to resist a change in pH as acid is added (by the bacterial processes in aquaponics). Buffering uses calcium carbonate, usually in the form of crushed oysters or eggshells, to increase the buffering capacity of the water. In practice, carbonates will keep water at a fairly stable pH of 7.0–7.5 even if excess carbonates are added. This method is forgiving of mistakes as excess dissolved carbonates will precipitate out of solution until the carbonate level in the water drops to the point at which the precipitated carbonates begin to slowly redissolve, thus continuing to buffer the water. When using the buffering method, supplemental potassium should be added regularly in the form of potassium sulfate (K_2SO_4), which has a fairly neutral pH.

The primary drawback to the buffering method is that the pH is maintained at around 7.0–7.5 which is not ideal for plant growth. As we prioritize crop production, this is not satisfactory.

The method of "adjusting" the pH uses calcium and potassium in hydroxide forms to incrementally adjust, rather than buffer, the pH. Hydroxides are more soluble in water than carbonates, and only very small amounts are needed to make adjustments. Hydroxides are also used up rapidly in this method, so adjustments need to be made regularly. Smaller and more frequent adjustments enable a much more precise level of pH control than is found in the buffering method, and the operator can keep the pH in a range that is more ideal for plant growth. If using this method, the use of both calcium hydroxide and potassium hydroxide will help to maintain the desired equilibrium of these minerals in the water.

IMPORTANT

Calcium and potassium hydroxides are very caustic and should only be handled while wearing personal protective equipment.

The drawback to the hydroxide method is that it is less forgiving of mistakes: if too much of either hydroxide is added, the pH will spike rapidly and possibly kill both fish and bacteria, and if not measured and adjusted regularly, the pH may quickly fall to a similarly toxic level.

We recommend using the hydroxide method, including an automated controller, to allow precise control of the pH. This enables far superior plant growth.

pH Management

Your system will likely only require pH adjustment up or down, most likely up. Most pH controllers can be set to adjust up or down but not both, which is fine. If your source water is particularly hard, typically due to a high level of calcium and/or magnesium, you may find that your system naturally has an ideal pH of 6.5 with little or no adjustment, or you may even find that you have to lower the pH. For most water sources, you will likely only ever be adjusting the pH up. Test your incoming source water before it enters the system. Initially, you will do this regularly. As you learn your system, this can be much less frequent.

Calcium and potassium must always be relatively equal in concentration within the system as they have a direct competitive relationship within a solution. If one of these elements is present at a significantly higher level than the other, the element of lower concentration will become insoluble and precipitate out of the solution, thus becoming unavailable for plant uptake. This typically only occurs in extreme cases, but to lower the risk, calcium and potassium are added in relatively equal amounts. Test for concentrations of each on a regular basis.

Calcium hydroxide is most commonly available in powder form and potassium hydroxide as a liquid, so they must be added to the system through separate means. Potassium is dosed via the pH controller and calcium via a sock in the CFB.

The pH controller probe is located near the pump intake in the sump. This tests the water after being mixed in the sump and before it enters the tanks. The pH dosing tube inputs at the opposite (east) end of the sump, near the water return from the hydroponic subsystem, to allow proper mixing before the water reaches the probe. The goal of inputting potassium hydroxide via the pH controller is to maintain a minimum pH of 6.5. Potassium hydroxide must be pure, meaning the liquid contains only potassium hydroxide and water. Check the MSDS to confirm. We use Advanced Nutrients pH Up.

Calcium hydroxide, also known as hydrated lime or agricultural lime, is almost never pure and typically contains some magnesium oxides which are also beneficial for plants. Always check the MSDS to ensure it does not contain any additives or anti-caking agents.

Put 2–3 cups of the calcium hydroxide into a sturdy fabric bag or sock, seal the sock and place it in the downstream (south) end of the CFB. Every second day, simply give the sock a quick shake. Though imprecise, we have found it to be the simplest and most consistent method of supplementing calcium in the system. Avoid shaking the sock too vigorously or you may add too much calcium and raise the pH higher than intended.

When first starting, do a chemical test for pH, calcium and potassium 4–6 hours after each sock shake. As you use this method and witness its effect on pH, you will quickly learn what is best for your system, and it will become second nature. The objective from a mineral standpoint is to keep at least 40 ppm of both potassium and calcium in the system.

The pH controller will cease dosing for a time after each sock shake as the pH rises. The job of the controller is to simply maintain the minimum pH at all times.

If pH down is required, there are several options, but we suggest phosphoric acid (H_3PO_4), as this will also add phosphorus to your system. Check the MSDS to confirm purity. We suggest Advanced Nutrients pH Down.

If your source water is hard due to high calcium levels, you will need to add potassium to keep the two reasonably balanced. We recommend supplementing with potassium sulfate (K_2SO_4).

Once you have learned what your system requires for pH management, if the pH becomes unstable (swinging more than 0.5 regularly), something is likely wrong within your system. Likely causes include a dead fish in the system or a process you have overlooked. Immediately do a full batch of chemical tests and stop feeding the fish until you discern the problem. Fish can survive for weeks without food. After approximately 24 hours, they go into a reduced energy state in which their metabolic processes greatly decrease, they consume very little oxygen and produce little to no ammonia. In most cases, you should be able to discern and resolve the problem within 24 hours.

Water Temperature

Our system is a cold-water design, most appropriate for temperate or colder climates. This is very well-suited to many types of plant and certain types of fish. It is less well-suited to both types of bacteria.

Many plants thrive in water that is 10–20°C (50–68°F), primarily leafy greens such as lettuce, kale, chard, the choi family, mustard, and herbs such as mint and watercress. Some plants that you might assume would thrive, such as spinach, grow poorly in a DWC system, as they do not like having their roots submerged in water. Most fruiting plants are not well-suited, as cold-water aquaponic systems are typically abundant in nitrogen but relatively deficient in the phosphorous required for fruits and flowers.

Cold-water fish have a narrower preferred temperature range of 12–17°C (54–63°F). Water above 17°C will dramatically slow growth and increase the risk of infection and disease. Warmer water also has a lower oxygen carrying capacity.

Bacteria are the crucial entities in determining water temperature. Both types prefer warmer water of at least 20°C. Below 20°C, growth and performance of the bacteria progressively drops. The optimum temperature is therefore as high as the fish can tolerate in order to minimize the loss of growth and performance of the bacteria.

IMPORTANT

The RCA system is designed to operate at 16°C (61°F), +/- 1°C.

Dissolved Oxygen (DO)

The amount of oxygen that can be dissolved in water is dependent on the temperature: colder water is able to hold more DO than warmer water. The maximum amount of oxygen that can be dissolved in water at 15°C (59°F) is about 10 ppm; at 25°C (77°F) it is only 8 ppm. This is a major advantage of cold-water aquaponics versus warm-water, which typically operates at around 25°C. The aerators used in our design keep the water saturated at all times.

The entities which have the most need of oxygen are the fish. Trout require at least 5 ppm to survive and 7–9 ppm to thrive. DO concentrations below 5 ppm will quickly suffocate trout. Plants also need continuous high levels of DO to thrive, although both the plants and bacteria can survive for short durations at concentrations as low as 1 ppm. All entities in the system thrive with higher concentrations of oxygen, thus we supply the maximum possible.

As fish need the highest levels of DO, the objective is for the water to be fully saturated just before it enters the tanks. For the first year of operation, test DO every day or two. Always test DO as close as possible to the pump intake in the sump and just after each tank by hanging the probe down the SPAs. The first measurement (in the sump) should always read 100% saturation (approximately 10 ppm). The second measurement (in the SPA) should be no lower than 4 ppm. The difference between the two readings is the amount of oxygen the fish consume. If the first reading shows full saturation but the second is lower than 4 ppm, either the fish biomass in that tank is too high, thus they are consuming more oxygen than the system provides, or the system flow rate is too low and thus oxygen is not being replaced in the tanks quickly enough.

After some time, testing can become less and less frequent. We rarely test for DO anymore except immediately after adding a new cohort and in the transportation of cohorts.

Optimum Water Quality Parameters

	Trout	Plants	Nitrifying Bacteria	Mineral Bacteria	Our System
pH	6.5–8.5	6.0–6.8	Prefer 7.2–8.0	Prefer 7.0 or higher	6.5
Temperature	12–17°C	10–20°C	15–35°C	15–35°C	16°C
Dissolved oxygen	9–10 ppm (Min 5 ppm)	9–10 ppm	Min 5 ppm	Min 5 ppm	9–10 ppm (saturation)
Maximum ammonia	3 ppm	10 ppm	10 ppm	3 ppm	Less than 1 ppm
Maximum nitrites	1 ppm	5 ppm	10 ppm	1 ppm	Less than .25 ppm
Maximum nitrates	800 ppm	1000s ppm	Unknown	Unknown	Less than 1000 ppm

7

Cycling the System

"Cycling" is the establishment of the biofilter (nitrifying bacteria) and, in large systems like this, the gradual introduction of fish and plants. Once the system is established with a robust and thriving biofilter and a full complement of fish cohorts and plants, you have successfully cycled your system.

Filling the System with Water

Before cycling the system, it will need to be filled with source water. As the cycling process is done in stages, at this time fill only Tank 1 and one trough. Turn the valves off to Tanks 2 & 3 and the other two troughs. They will be brought online when you introduce the second and third cohorts.

Fill both sides of the trough at the same time to keep the liner from shifting. Fill the sump, Tank 1, the Tank Manifold, RFS and CFB. During this process, check every visible connection for leaks and confirm all equipment functions properly.

Once the system is full (one tank and one trough only), turn on and test both pumps, both aerators and the UV sterilizers.

If the source water is chlorinated, you will need to let it circulate for 24 hours with UV turned on to remove the chlorine before beginning the cycling process. When you later fill the second and third sets of tank and troughs, let them sit for 24 hours before bringing them online. Once online, allow the system to circulate for a few hours to mix before inputting fish.

During the cycling process, the UV sterilizers are turned off until the biofilter is established. For the first cohort of fish, turn on two of the four units. Close the flow to the other two units. Add the third unit when the second cohort of fish is input. Add the fourth unit when the third cohort is input.

IMPORTANT

Two to three weeks before beginning the cycling process, begin germinating the first batch of seeds. Move them to the seedling area after one week and start new batches weekly. There is no way of knowing exactly how long the cycling process will take, but you must have seedlings ready to plant once the biofilter is generating nitrates. The goal is to have mature seedlings ready to transplant into a trough once the nitrate concentration has risen above 50 ppm and ammonia/nitrite levels are low. If seedlings outgrow their trays before the system is ready for them to be transplanted, consider it practice and compost them. As you will be planting only ¼ of one trough initially, only five trays of seeds should be needed per week. You will need to feed seedlings with a store-bought fertilizer until your system water contains the required minerals. See Chapter 9 for instructions on germination and seedlings.

Cycling the System

This is it: the moment when you start creating a living ecosystem. You now have a fully functional system of inanimate objects (troughs, tanks, plumbing, etc.). Water flows in a complete loop and is oxygen saturated by the aerator. It's time to cycle the system into life.

The first step is to start establishing the biofilter by adding a source of ammonia and introducing nitrifying bacteria. The second step is testing water quality to chart the growth of the biofilter and confirm it is functioning. The final step is the introduction of fish and plants in stages over 9–12 months until the system is at full capacity.

Ammonia is the energy source for nitrifying bacteria and at this point is the only element missing. The first decision is to choose a source of ammonia to input. There are two options: add pure ammonia directly into the water or use fish to generate ammonia. We strongly recommend adding pure ammonia. This allows precise control of ammonia levels and faster startup times as higher levels are possible. Using fish to generate ammonia at this stage has

IMPORTANT

Do not rush this process. The biofilter grows slowly, and introducing fish too soon will cause ammonia spikes and failures. We learned this lesson the hard way.

two major problems: the fish are very likely to suffer or die, and the ammonia levels produced are not directly controllable by you. For reasons of both ethics and efficiency, using fish initially does not make sense unless a pure source of ammonia is not readily available.

If it is necessary to cycle the system with fish, use feeder goldfish from the pet store and consider them sacrificial. You should only need 0.5–1 kg of goldfish. Use small amounts of feed initially until you have an idea of how much ammonia they are producing.

IMPORTANT

When adding ammonia directly, it is mandatory to use only pure ammonia. Many common brands contain fillers or other chemicals, such as perfumes, foaming agents or additives. As a rule of thumb, if the product is scented, foams when shaken or is not clear, do not use it. Always refer to the MSDS to ensure that the product contains nothing but pure ammonia and water.

There are two options for establishing the biofilter: seed the system with store-bought bacteria cultures or allow the bacteria that are naturally present in the environment to create colonies. The potential advantage to the latter is that the bacteria will be more locally adapted. The disadvantage is that it will likely take weeks longer than seeding the system. Without seeding, establishing the biofilter will likely take 1–2 months; with seeding, it may only take 2–3 weeks. If you use a seed bacteria, it must be approved for use in food fish systems. We used and recommend Proline Nitrifying Bacteria.

Don't add ammonia or seed bacteria yet.

IMPORTANT

The UV units must be turned off for the duration of the biofilter establishment.

Take baseline measurements of ammonia, nitrites, nitrates, calcium and hardness. Record the concentrations of each element in a log book. Ammonia, nitrite and nitrate levels should all be 0 ppm, calcium above 40 ppm, and the pH should be between 6.5 and 8.0. If either calcium or the pH is too low, increase by mixing small amounts of calcium hydroxide into the sump. If the pH is too high, lower by mixing small amounts of phosphoric acid into the sump. After mixing additions into the sump, wait for the additions to mix in the system before retesting.

Slowly add 2–3 ppm of ammonia to the system, adding 1 ppm at a time and testing to confirm the concentration. To determine how much ammonia will equate to 1 ppm, take the total volume of water, then calculate the weight of ammonia required. 1 ppm is equal to 1 milligram per liter. For the RCA system, the total volume of water is approximately 66 cubic meters (66,000 liters); therefore, 1 ppm equals roughly 66g of ammonia in a full system.

Once you have sourced the ammonia, add 66g to the system. If your ammonia is in liquid form, it may contain water, so you will need to calculate the equivalent of 66g (e.g., a 50% water/ammonia solution will require adding 132g total). In order to mix throughout the system more rapidly, add the ammonia at numerous locations.

After several hours, test and log the ammonia level. If your baseline was 0 ppm and is now 1 ppm, you've added the perfect amount. If not, adjust the dose accordingly. Continue to add ammonia in this manner until a concentration of 2–3 ppm is reached. If too much ammonia is added (> 3 ppm), replace system water with source water until the level is less than 3 ppm. (If you are using goldfish instead of adding ammonia directly, feed them small amounts and test the ammonia level often. Aim for a target ammonia concentration of 1–2 ppm. This may take a day or two to achieve.)

Wait several hours and test the ammonia level again, then several more hours and test again. Once the level has stabilized between 2–3 ppm, pour in 4 liters of Proline Nitrifying Bacteria in the most upstream (north) end of the CFB. If you are allowing the naturally occurring nitrifying bacteria to colonize the system, all you have to do is wait. Wherever ammonia is present, ammonia oxidizing bacteria will soon be found.

The objective now is to continuously feed the bacteria the ammonia they require to exponentially multiply. Maintain an ammonia concentration of 1–2 ppm at all times. Twice per day, test for ammonia, nitrites, nitrates, calcium and pH. Create a chart for these tests to gauge progress over time. If calcium levels drop below 40 ppm or the pH drops below 6.5, increase with small additions of calcium hydroxide. You do not need to add potassium at this stage.

The first indication of biofilter establishment is a drop in ammonia and rise in nitrites as the nitrifying bacteria start the first stage of oxidation. When ammonia levels drop below 1 ppm, add 1 mg of ammonia for every liter of water in the system. Always add ammonia in various locations to give it ample time to mix prior to reaching the primary bacteria colonies in the CFB. The objective is to keep the ammonia between 1–2 ppm until the biofilter is fully established.

It may take one week or more to see the drop in ammonia and increase in nitrites. Continue adding ammonia as needed and testing/charting twice daily.

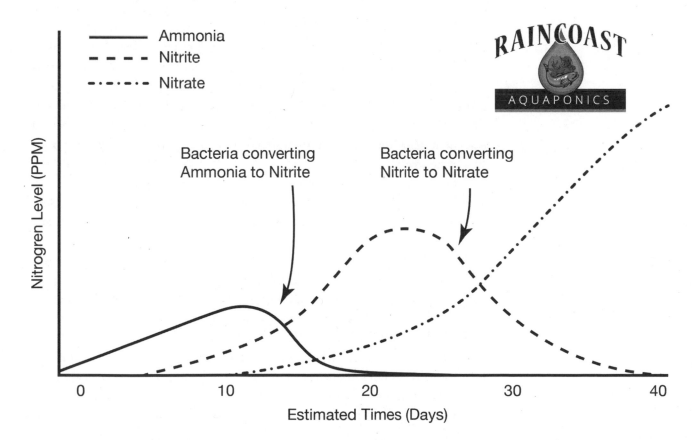

The next indication of successful cycling is a slow decrease in nitrites and increase in nitrates as the nitrifying bacteria convert nitrites into nitrates during the second stage of oxidation. Continue adding ammonia and testing/charting twice daily. Add calcium as needed to keep above 40 ppm at all times.

The final indication is a radical drop in both ammonia and nitrites and a consistently high level of nitrates as the bacteria become established sufficiently to convert ammonia rapidly to nitrites and then almost instantly to nitrates. Continue adding ammonia and testing/charting twice daily for at least one week to confirm that the biofilter is stable and successfully converting ammonia to nitrates rapidly. At this point, you should be reading ~1 ppm of ammonia and <0.25 ppm of nitrites when measuring at the sump.

You have now established the biofilter. Continue adding ammonia to feed the biofilter until ammonia additions are replaced by the first cohort of fish.

In a smaller system, the next step would be to add fish and plants up to the system's capacity. In a system of this size with multiple fish cohorts, fish and plants are added gradually. It takes up to a year to bring the system to 100% capacity (assuming no major problems).

The Importance of Calcium in Cycling

Our source water is very soft and lacks calcium because it does not percolate through limestone on its journey into the local aquifer. When we first cycled our system, we were unaware of the vital importance of calcium in the cycling process. We added ammonia and seed bacteria, tested and charted, and saw the expected indicators. Sadly, we proceeded to add the first fish cohort and lost them when the bacteria ran out of calcium and crashed, leading to a massive spike in ammonia.

Calcium is a necessary element for bacteria to function. If your water contains less than 40 ppm, supplementing is required before seed bacteria is added.

We recommend using calcium hydroxide. Use with caution to avoid raising the pH too high. It is important to keep the pH below 8.0 and ideally below 7.0 during the cycling process. Different bacteria thrive at different pH levels. Your system will be operating at 6.5, so you want to establish bacteria that will thrive at that pH.

The First Cohort of Fish

Start with a half cohort of fish initially so as not to overwhelm the biofilter. For the RCA system as per the calculations in Chapter 3, this means approximately 125 fish that weigh 10–20g each for a total initial biomass of 1.25–2.5 kg. The second and third cohorts can be full cohorts.

A few days before adding the fish, stop adding ammonia and wait for the concentration to drop below 1 ppm before introducing them. You should never have to add ammonia to the system again as the fish will now create it naturally.

Turn on two of the UV sterilizers. The UV units will now always be running during normal operation.

For the first 2–3 days after introducing the first cohort, give only ¼ of the calculated daily feed. Increase the amount gradually over 1–2 weeks to reach full daily feed rate. Repeat this process for all future cohorts. See Chapter 8 for details on introducing new cohorts and calculating the daily feed rate.

Continue testing and charting water quality and pH daily, collecting samples from a tank inlet spout. It is very important once fish are in the system that the ammonia level does not exceed 3 ppm and nitrites 1 ppm. If your biofilter is growing quickly, this will not be a problem. If you see results at or close to either of these levels, immediately stop feeding the fish until the levels reduce, then gradually re-introduce feed.

The First Year of Operation

Since before the start of the cycling process, you have been germinating five trays per week, so you should now have large healthy seedlings, ready for transplant. Allow the nitrate concentration to rise above 50 ppm before transplanting the first seedlings. At this time, begin supplementing iron and magnesium (see Chapter 6). At this point, the pH controller should be online and you should be shaking the calcium sock gently and infrequently.

During the first four months of operation, you will have only a half cohort (125 fish). Once nitrates are above 50 ppm, plant ¼ of the first trough (22 rafts) with the largest, healthiest seedlings. Replant the first ¼ trough whenever the crop matures and is harvested. So long as there are no signs of nitrogen deficiency, plant another ¼ of the trough one month after the first (one side is now fully planted), then add the third ¼ at the start of the third month and the fourth ¼ at the start of the fourth month. By the fourth month, the first trough should be full at all times.

For the first four months, seeding five trays should provide ample seedlings, though you may require more depending on the production times. If you have excess seedlings, consider it practice and plant only the largest and healthiest. After four months, you should have one trough fully planted at all times, and you are now ready for the second cohort.

Prior to introducing the second cohort, fill Tank 2 and the second trough. Allow any chlorine to dissipate for at least 24 hours before bringing the second tank and trough online. Once online, let the system circulate for a few hours prior to inputting the second cohort.

If you are rotating cohorts between tanks, transfer the first cohort to Tank 2, sampling and counting the fish as you move them (see Chapter 8). Input the second, full cohort (250 fish) and follow the same ¼ planting progression in the second trough, adding ¼ of the trough per month. Seed ten trays per week from now on. Eight months after introducing the first cohort, you should have two full troughs at all times.

Follow the same pattern for the third cohort (250 fish) and trough. Seed 15 trays per week from now on.

Throughout the first year, continue testing and charting ammonia, nitrites, nitrates and pH daily to monitor the biofilter. As the biofilter matures and stabilizes, and you become familiar with your system, testing frequency can decrease to weekly. If the plants show signs of nitrate deficiency, decrease the planting progression speed or increase the daily feed rate. Test weekly for calcium, potassium and iron, and perform regular plant health observations (see Chapter 9).

Full Capacity

Once you have three cohorts (including the first half cohort) and three full troughs, you have successfully cycled your system to full capacity. This will take one year, assuming no major problems or losses of cohorts. Now a robust and symbiotic ecosystem, your system will continue to establish itself and stabilize for some time. When you harvest the first cohort, rotate the fish tanks and introduce a full fourth cohort. You do not need to follow the same gradual ¼ trough planting progression again. At this point, you should be harvesting, transplanting and seeding at least 15 trays per week (see Chapter 9). Strive to have all troughs full at all times.

Raising Fish

Fish Species

THERE ARE SEVERAL POTENTIAL OPTIONS FOR SPECIES OF FISH to raise in a cold-water aquaponic system. The only mandatory characteristic is that they grow well in cold water (under 18°C/64°F).

Since day one, we have raised *Oncorhynchus mykiss*, commonly known as rainbow trout, or just "trout" in this book. Other species, notably sturgeon and coho (salmon), may also be well-suited to cold-water aquaponics but to date we have not experimented with them, so we cannot speak from experience.

Trout have numerous advantages that make them ideal for aquaponic production:

- Fingerlings are usually easy to source in temperate regions.
- They typically require little licensing or oversight by governing bodies.
- They have an excellent feed conversion ratio.
- They are well-known and always in demand, thus easy to market.
- They are tolerant of the lower pH (6.5) which is required for optimum plant production.
- Their ideal water temperature is 15–16°C (59–61°F), perfect for cold-water aquaponics.

Feed Conversion Ratio

Feed Conversion Ratio is the capacity of fish to convert feed into body mass — the lower the ratio, the better. You may read that trout have a ratio of as low as 1.1, which means that for every 1.1 units of food weight, the fish will gain 1 unit of biomass. In our experience, this holds true only under perfect conditions, so expect to see a slightly higher ratio.

Fish Sourcing

You will most likely be sourcing fish from an external source, as operating a hatchery is a completely separate process not covered in this book and likely not something you should consider unless you are experienced in aquaculture. The ease of accessing a fish supply will depend greatly upon your location.

Seek a hatchery that is Certified Disease Free by your local fisheries authority and that can provide a Certificate of Fish Health for fingerlings you purchase. Note that even Certified Disease Free hatcheries will occasionally encounter disease problems, but a good hatchery will remedy these quickly as they arise and problems, should not be ongoing.

The size of the fish you put into your system will largely be determined by the availability and cost of fingerlings in your area. The amount of time they spend in the system is determined by their growth rate and target harvest weight.

We sell almost all our fish to a local restaurant that requires them to be minimum 1 kg. Accordingly, we source fingerlings which are typically 10–12 cm long, weigh 10–20 g and yield average 1 kg (or larger) fish in 11–12 months. If you want larger fish at harvest, you can either source larger fingerlings or leave them in the system longer. Use the formulas in Chapter 3 to determine the correct number of fish and their expected growth rate.

Fish that are smaller than fingerlings are called "fry" and must be fed a special crumble feed. We do not recommend them. Fingerlings are at least 3–4 months old and can eat pellet feed.

Always inspect fish at the hatchery or, if delivered, when they arrive. Check for weak or dead fish. Weak fish will often be floundering near the surface, lethargic and/or floating on their side or back. Check also for any obvious sign of disease. The most obvious will be patchy discolored skin or frayed white fins. Transporting fish can be stressful to them, so it is not unusual to see a few of them showing signs of weakness/lethargy, but there should be no dead fish. If there are signs of disease or more than a few dead fish, contact the supplier and have them resolve the problem or find a new supplier.

We pay approximately $1.00 per fingerling.

Record Keeping

Every cohort of fish grown in the system must be accompanied by its own log to track fish growth and health. In some cases, this may also be a legal requirement of having an aquaculture license. Assign each cohort a sequential number. This number will be used on the Fish Sample Log and the Cohort Log.

Periodic sampling of each cohort is recorded on the Fish Sample Log, and the data is used to calculate the biomass of each cohort. The Cohort Log is used to track the growth of the cohort over time, as well as the number of fish and total biomass. The data is used to calculate the daily feed rate for each tank, and the rate is recorded on the Daily Log.

See Chapter 11 for sample logs.

IMPORTANT

The total biomass in the system is calculated by adding the total biomass of each tank as recorded on the Cohort Logs, in order to ensure your system is operating within the target biomass range as determined in Chapter 3, and to determine when to harvest fish and how many fish to harvest.

Fish Transport

If you are picking up from a supplier, you will need a live-hauling setup with a tank, oxygen supply and a DO meter. Live-hauling fish means building a portable life-support system. There are several considerations to take into account, notably stocking density, temperature, DO level and water quality. During transport, the conditions must be kept at or better than the standard environmental conditions for trout: 15–17°C (59–63°F) and more than 7 ppm of DO The stocking density can be higher than normal as long as care is taken to prevent excessive ammonia or solids buildup and the DO level is maintained.

If you are following our design, you will be inputting cohorts of 250 fish that are between 10–20g each (2.5–5 kg biomass). A biomass of 5 kg could potentially be transported by car in a large cooler (> 50 L), but this does not leave much wiggle room and can be undersized if the fish are on the larger end of the scale.

We recommend hauling by truck and either buying an insulated fish tote or building a custom hauling tank. We recommend the tank be at least 160 L to accommodate up to 10 kg biomass for transport that is four hours or less.

Our tank is 3′×3′ by 1.5′ high, holds approximately 380 L, weighs 380 kg (840 lbs) when full and was sized to fit our small farm truck. With this tank, we are capable of hauling a maximum of 35 kg of fish using supplemental oxygen. If you build a custom tank, it should be constructed out of ¾″ marine-grade plywood and waterproofed with 2 or 3 coats of epoxy resin. See IMPORTANT note on Epoxy and Fish Safety in Chapter 4.

The tank must be well constructed and rugged enough to handle the forces of cornering and braking when loaded with water. Make sure to include heavy-duty tie-down rings.

The tank must have a watertight lid and a drain valve on one side. Construct the lid to overhang all four sides and use a foam gasket to create a watertight seal. Install a bulkhead fitting in the center of the lid with a standpipe that rises 6". When hauling, run the oxygen line and DO probe through it. You can also use it to top up the tank if needed.

Dissolved oxygen, temperature and time are the biggest considerations for transport. DO is by far the most important. Once the fish are in the tank, they will immediately begin consuming oxygen, and if they have been fed in

IMPORTANT

For biosecurity, always sterilize the hauling tank prior to and after every transport. Use a germicidal cleaner and wash the tank off-site, preferably at a car wash.

Live haul setup: fish transport tank, oxygen tank, pressure regulator, flow meter and micro-bubble diffuser.

the past 24 hours, will also be producing ammonia. The longer the trip, the more ammonia buildup and oxygen consumption. These can be decreased by withholding feed 24–48 hours prior to transport, enabling longer trips or more biomass per trip.

Unless you are using an insulated hauling tank or making very short trips, only transport fish when the temperature is cool. During the summer, transport at night or early in the morning.

Oxygen supplementation is mandatory during transport, even for small distances. Fish will consume the available oxygen surprisingly fast, if it is not replaced. Use a portable oxygen tank specifically for transport. Connect the tank to a single Sweetwater Micro Bubble Diffuser via a flow meter and regulator installed through the port in the tank lid. To minimize stress, DO levels should be maintained at 100–150% saturation during transport.

Bring your DO meter with you for transport. If it is possible to have the probe in the tank and the meter in the cab of the truck (running through a window) this is ideal, as you can monitor oxygen levels at all times.

Additionally, you can add 0.5% salt (5g salt per liter of water) to the tank before transport. This serves as a prophylactic pathogen treatment, increases fish tolerance of ammonia and nitrites and helps to reduce stress. The salt must be pure sodium chloride (see Salt Bath Treatment later this chapter).

IMPORTANT

Always fill the hauling tank completely full of water, even if only a small number of fish are being transported. If an air gap is left at the top of the tank, the water will slosh around violently. This "free surface effect" will stress and damage the fish, and movement of water in the tank can destabilize the vehicle enough to cause it to roll over. When the tank is full to the lid with no air gaps, the liquid mass behaves as though it was a solid and maintains a stable center of gravity when the vehicle is cornering or leaning.

Fish Tempering and Quarantine

Regardless of how fish arrive at your farm, follow these procedures for tempering and quarantine. Before receiving fish, make sure to fill the Purge Tank half full with system water and turn on a circulation pump or supplemental oxygen to prepare for quarantine.

Tempering

As a rule of thumb, fish should not be exposed to a temperature change of more than 5°C or a pH change of 1.0 in a 20-minute period.

Immediately upon arrival, test the hauling tank water and your system water for temperature and pH. If there is a difference between them of more than 5°C or 1.0 pH, you must temper the fish in the hauling tank prior to moving them into quarantine in the Purge Tank. Add system water to the hauling tank, exchanging no more than 10% of the tank volume every 10–20 minutes (slower is more gentle on the fish). Ice can be used to chill the hauling tank if needed on hot days.

Once the water in the hauling tank is within 5°C and 1.0 pH of the system water, the fish can be quickly and gently moved into quarantine in the Purge Tank. Use the dedicated Purge Tank fishing net.

Quarantine

The fish will typically remain quarantined in the Purge Tank for 1–2 weeks. The purpose of the quarantine is to avoid spreading pathogens or disease to other cohorts. The UV sterilizers will greatly assist in preventing this, but all due care should be taken. The introduction of a new cohort from an external source is the most high-risk moment for pathogens to enter the system.

IMPORTANT

The Purge Tank must be isolated until the end of quarantine (not connected to the main system). Supplemental oxygen must be added by either the backup oxygen system or with the use of a submersible circulation pump to agitate and aerate the water.

As the Purge Tank is not connected to the main system, including the biofilter, the ammonia level will slowly increase. Test ammonia at least once per day. When it reaches 1 ppm, remove half the water and replace it with system water. Do this as often as needed.

IMPORTANT

Water from an external source, such as the hatchery, and water used during quarantine must never be added back to the main system, the Waste Tanks or a natural body of water (pond, stream, etc.). It should also not be used to water soil crops. It can be used to water trees or must be discarded in a field or sewer.

Observe fish regularly for signs of disease. Regardless of whether anything is seen to be problematic, immediately before moving the fish to Tank 1, give them a 1% Salt Bath Treatment for 2–3 hours.

Label for Bio-Oregon feed.

Fish Feed

Fish feed is a contentious issue, particularly for carnivorous species. Carnivorous fish typically eat other fish. The notion of feeding wild fish to farmed fish is condemned by some. We acknowledge these issues but will not be addressing them in this book. We strive to use only the most environmentally friendly, ethical feed available and hope you will do the same.

Trout are carnivorous and require a high percentage of protein as well as essential oils and fatty acids. The quality of feed varies greatly, particularly in concentrations of oils and lipids. We use and recommend Bio-Oregon feed, made by Skretting, which consists of by-product from marine sources. Like all fish feeds, the content changes regularly based on the supply available to the company. This feed is "Best Aquaculture Practices Certified" by the Global Aquaculture Alliance.

The protein alternative to marine-sourced feed is land-sourced, typically from animal sources such as chicken or vegetable sources such as soy or corn. Without needing to consider the huge

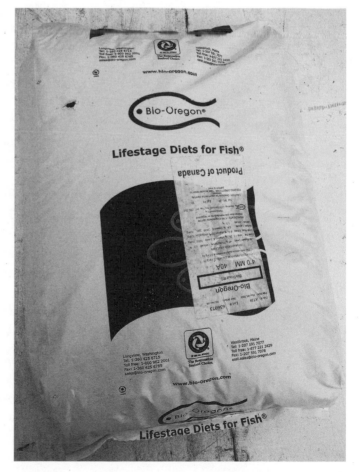

environmental and ethical problems with factory-farmed animals and mono-crops (which is how virtually all soy and corn is grown), land-based feeds are generally less desirable because while they can meet crude protein needs, the quality of the feed is generally inferior and lacks the needed oils and lipids.

Feeding Your Fish

The amount of daily feed is based on the biomass (total weight of fish) in each tank, and thus increases as the fish grow. Keep track of the biomass by conducting regular sampling of the tanks and recording results on the Cohort Logs.

Sampling

Sampling is typically done every three weeks by weighing a few samples of one cohort and using that data to estimate the average weight per fish. Sampling data is recorded in the Fish Sample Log, and the average weight per fish is then recorded on the Cohort Log for the appropriate cohort. The Cohort Logs are used to track the growth rate and biomass of each cohort and to determine the daily feed requirement.

1. Place a tote of water with a lid on your large scale and tare the weight.
2. Scoop a partial net-full of fish from the tank using the dedicated tank net and place them in the tote with the lid on to prevent water splashing out.
3. Wait a few seconds for the scale to settle and record the weight of the sample on Fish Sample Log.
4. Count the fish as you return them to the tank and record on Fish Sample Log.
5. Divide the weight of the sample by the number of fish in the sample to find the average weight per fish. Record on Fish Sample Log.
6. Repeat steps 1–5 three more times, recording the data from each sample.
7. Add together the average weight per fish from each of the samples and divide this by 4 to find the mean average weight per fish. Record on Fish Sample Log.
8. Record the mean average weight per fish on Cohort Log, noting "sample" in "Event" column.

Your feed manufacturer will have a chart of the recommended daily feed rate that shows the amount of feed to give as a percentage of body weight for a given water temperature and size of fish. In general, for trout in 15°C water, younger fish are fed approximately 2% of their body weight per day until they reach approximately 100g, after which feed is reduced to ~1.5% of body weight per day, and then further reduced to ~1% beyond 500g. Smaller pellets (3–4mm) are fed to the youngest fish, and as they grow, the pellet size is increased (up to 9mm for the largest fish we grow).

Feeding Technique

It is better for the biofilter if feed is divided throughout the day rather than given all at once, as that will create a spike in ammonia. The primary aim of the aquaponic farmer is system stability, so doing this is undesirable. While it is acceptable if necessary to input all feed at once, strive to divide feed over two or three feedings throughout the day.

To feed by hand, simply scatter the feed onto the water surface. It will naturally start sinking slowly and should cause a feeding frenzy where fish jump and compete for food. A frenzy is normal and an indication of good fish health. If the fish do not frenzy, this is often the first clue that something is wrong.

A feeding frenzy is a sign of healthy fish.

If you choose to use auto-feeders, you will require one for each tank. The advantage to auto-feeders is that they will divide the feed throughout the day, far better than is possible by hand. Depending on the size of the feeder, they will allow you not to be onsite for a period of time. The disadvantage is that you will be less intimate with the behavior of the fish. The act of hand feeding daily and witnessing their state when fed is extremely useful in gauging fish health and vitality. An auto-feeder typically costs $300–1,000.

We recommend feeding by hand with feed divided over 2 or 3 times per day. If you use an auto-feeder, select high-quality belt feeders and avoid demand feeders which let the fish dispense the feed on demand.

Fish Tank Rotation

You have two options for raising fish: raise them in one tank for their life (after quarantine) or rotate them from tank to tank as cohorts are harvested. The advantage to using a single tank is that the rotation process (netting the fish from one tank to another) is stressful to fish. The advantage to rotation is that you can more closely monitor the health and number of fish and you can easily remove runts from the cohort.

This is a personal choice, as both options are acceptable. We rotate our fish, as we feel the opportunity to observe and count them outweighs their brief stress. Rotation also allows us to easily perform a prophylactic Salt Bath Treatment and gives us the opportunity to fully clean and sterilize each tank on a regular basis.

When transferring between tanks, carefully count the fish as you deposit them into their new tank and record on the Cohort Log. Use the opportunity to sample the cohort and update the total biomass on the Cohort Log. It may be necessary to partially drain the tank (store the water in the cistern) in order to more easily catch the last few.

Tank Cleaning and Sterilizing

As fish tanks are open to light, algae will grow prolifically on the sides and bottom even if the tanks are covered with a shade cloth or other fabric. If allowed to build up, algae can become detrimental to the system. Thick layers will consume oxygen as they grow and die and can harbor potentially pathogenic fungal spores and bacteria. It is important to remove algae regularly.

Every week, use the dedicated brushes for Tanks 1–3 to scrub the walls and floor of one tank, rotating weekly so each is scrubbed every three weeks. If the algae grow back quickly, you may need to clean more frequently.

Scrubbing can be done with fish in the tank — there is no need to remove them or drain the tank. If there are large amounts of algae, the cleaning may need to be done in stages to prevent the water from getting too murky. Be sure to scrub the drain screen and the drain sump, if the tank has one.

Sterilize a tank whenever it is empty of fish, such as when it is fully harvested or the fish are being rotated between tanks. Empty the tank of water (move to the cistern) and scrub all algae from the walls, floor and drain. Rinse the tank into the waste collection system by popping the SPA.

In a clean bucket, mix 15–20 mL of 30% hydrogen peroxide per liter of water to scrub the tank. Use only the dedicated tank brush. Let stand for an hour before rinsing well. At the same time, sterilize the dedicated tank net and the dedicated scrub brush by standing them in the bucket of H_2O_2 solution. Rinse well before standing them to dry. The tank cover can be sterilized by submerging it in a bin of the same H_2O_2 concentration for an hour. Rinse before drying.

Top:
*Freshly cleaned fish.
Note the healthy pink
color.*

Center left:
*Our two-sided troughs
in full production.*

Bottom Left:
*Adrian harvesting
another great crop.*

Above: Our farm at sunrise.

Right: Young romaine enjoying the breeze through the roll-up side.

Below: True watercress, Nasturtium Officinale.

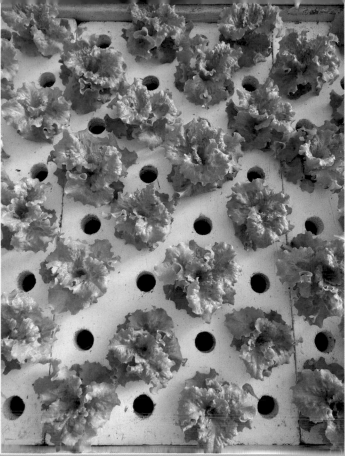

Tropicana green leaf lettuce, just after spacing/thinning.

Breen red romaine lettuce.

Left: *Alkindus red butter lettuce.*

Right: *Full troughs showing multiple stages of plant growth.*

Please keep in mind that this key is very general and not crop specific.
(Symptoms can vary quite a bit from crop to crop and from situation to situation.)

1

Is your plant Chlorotic?
(Is there yellowing of the leaves?)

YES   **NO**

a

- Is the yellowing only between the veins, with the veins remaining green?
- Is the young growth most affected?

If yes, then you most likely have an iron deficiency. If no, go to 1b.

New growth   Old growth

b

- Is the yellowing across the entire leaf (veins included)?
- Does the old growth appear to be more affected than the young?

If yes, then you most likely have a nitrogen deficiency. If no, go to 1c.

New growth   Old growth

c

- Is the yellowing primarily between the leaf veins, but the old growth is most affected?
- Are leaves cupping, developing brown spots, or dry with dead edges?
- Are the roots not developing? (Are there signs of stunted root growth?)
- Is there some chlorosis or dead fringes around the edges of the older leaves?

If yes, then you most likely have a potassium deficiency. If no, go to 1d.

New growth   Old growth

d

- Is the yellowing primarily between leaf veins, but the chlorosis is concentrated in older leaves, with the most affected leaves falling off (old chlorotic leaves falling off)?

If there are dead fringes along the leaf edges or dead, brown spots on the leaves and the old, chlorotic leaves fall off, then you likely have a magnesium deficiency.

Old growth

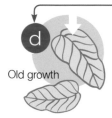

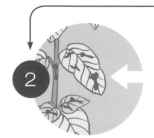

2

Are there necrotic spots (dead, brown spots) on the leaves or stems of the young, rapidly growing parts of the plant? Are there brown dead spots on the fruit (especially on the blossom end)?

If so, you likely have a Calcium deficiency. If no, then you may have another nutrient deficiency, but it is not as likely as other variables, including problems with temperature, pests, or cultural methods.

HOW TO TREAT THESE DEFICIENCIES

Iron
Add chelated iron to your system- there are many types out there, but FeEDDHA and FeDTPA are the best types of chelated iron. (FeEDDHA will turn your water red though.) There is quite a bit of math involved in determining how much to add, because different chelated iron types have different purities. Use the Able.ag Iron Calculator to easily calculate and schedule your iron additions.

Nitrogen
You should know that this is a possibility if you haven't been feeding much, have too much vegetation to fish, or have been measuring low nitrate levels in your system. Switch to a higher protein feed and feed more often to correct. If this doesn't correct, remove some plants and look for straw, wood or another high carbon substance in your system that could be consuming nitrogen in decomposition. Remove all wood, straw, etc. Low temperatures can also depress feeding and lead to nitrogen deficiency.

Potassium
In systems with low pH, add potassium hydroxide (caustic lye) to raise pH and supplement potassium. In systems with neutral or high pH, add kelp meal concentrate (0-0-10) or potassium sulfate (0-0-50) in very low quantities. In hydroponics potassium deficiences are very rare.

Magnesium
In systems with low pH, adding dolomitic lime can help, as well as hydrated lime. In both hydroponic and aquaponic systems, Epsom salts (magnesium sulfate) can be added in small quantities to supplement magnesium. In hydroponics magnesium deficiences are very rare.

Calcium
Most systems have plenty of calcium because it's common in the water. However, if there's too much potassium in the system, your plants might show a calcium deficiency. To correct in low pH systems, reduce the amount of potassium you are supplementing and add hydrated lime to the system in small quantities. In neutral or high pH systems, the best way to supplement calcium is with small amounts of calcium chloride applied foliage. In hydroponics calcium deficiences are very rare.

Top: *Freshly harvested watercress (left),
cut with roots for bulk sales (center),
and cut for salad mix (right).*

Above left: *Romaine hearts.*

Right: *Harvested Giant Red mustard (left)
and Abundance baby kale (right).*

Left:
*Freshly made
salad mix.*

Below:
*Layout of fish tanks,
Tank Manifold, RFS
and CFB.*

Above:
*Mature rainbow trout
raised on our farm.*

Right center:
*Fingerling with Bacterial
Cold Water Disease.
Note the skin lesion
and eroded fins.*

Below:
*Substrates from left to right:
loose rockwool, Pro-Mix,
expanded clay pebbles,
coco coir.*

Fish Health

To fully discuss fish health including diseases, symptoms and pathogens would require a voluminous scientific book. We will only scratch the surface in this brief discussion.

Fish health management consists of practices such as daily observation, record keeping, management plans and biosecurity protocols for the prevention and treatment of diseases. Observations and record keeping will help you to identify and treat potential problems early on, before they become catastrophic. Having a treatment plan is particularly important for aquaponic systems, as there are few treatment options available that can be used without harming plants and/or the biofilter. Below are two treatment protocols that we have found particularly useful for the pathogens we most commonly encounter.

Biosecurity protocols are used to prevent the introduction of pathogens into the greenhouse as well as prevent the spread of pathogens between cohorts. Biosecurity protocols are an integral part of the system and should be planned during the design phase. If you visit a hatchery, it will likely employ considerable biosecurity, and this is a good measure of the quality of the facility.

At a minimum, we recommend employing the following common sense biosecurity protocols:

- No outside water enters the system (only your source water).
- Monitor fish closely and regularly.
- No outside fish enter the system (e.g., from a pond).
- Fish are always quarantined and salt bathed prior to entering system.
- Feed is properly stored to ensure no mold.
- Visitors do not put anything or any part of their body in system water.
- Farmers must wash hands with biodegradable unscented soap before putting hands in system water.
- Use dedicated fishing nets and cleaning brushes for each tank.
- Install foot baths at all entrances (optional).

The most important methods of gauging fish health are experience and being intimate with the normal behavior of your fish. Given time, you will know how fish look and act when healthy. By observing them closely, you will notice anything out of the ordinary. This is a primary reason why we prefer and recommend feeding by hand.

Observe the tanks daily for signs of dead or weak fish. Dead fish will first sink to the bottom and if not promptly removed will begin to float as they start to decompose internally. Remove them immediately, then sterilize the net used with hydrogen peroxide or formalin. Weak fish will likely first be noticed floating near the surface and/or showing signs of lethargy and/or swimming irregularly. Remove weak fish immediately, as they are more likely to host disease or pathogens which can infect the entire cohort or system. While it may seem cruel to destroy a living fish for being weak, you must consider what it best for the cohort. If you ever lose a whole cohort, you will understand the concept of the greater good. Take mortalities offsite for composting.

It is normal to lose a few fish over time in your farm. You are creating an ecosystem where there is always variability of the conditions. In general, we consider losses of up to 5% of a cohort over a 12-month period to be within the range of normal. Some of these will be weak or stunted fish that are intentionally culled, other losses can occur from predation or jumping out of the tank. If several fish die or become lethargic at one time, it is a clear sign of a problem that requires immediate corrective action.

Pathogens are microorganisms that cause disease. They are highly opportunistic and will always be around your system but for the most part are kept in check by the UV sterilizers, good biosecurity practices and raising healthy fish. The number one cause of a pathogen or disease taking hold is

One of our early cohort losses. This is what you are trying to avoid at all costs.

fish stress. As in humans, there are many precursors to stress. The most common are instability in system pH or temperature, low oxygen, high ammonia or nitrites, physical damage due to overcrowding, greatly mismatched fish sizes in the same tank and moving fish from one tank to another. Your job is to keep the system stable and minimize stress to the fish however possible.

The most common pathogen we encounter is Flavobacterium, which can cause several diseases, most commonly Bacterial Cold Water Disease. Symptoms include lethargy and irregular swimming, followed by fin rot and skin lesions and eventually death. Fin rot is identified by whiteness around the fins and frayed, eroded or completely missing fins. Skin lesions begin as patches of discoloration and progress to open sores if untreated. See color photo.

If we witness a low level mortality problem (1–5 fish dying per day) or notice another sustained symptom for longer than a few days that is not directly attributable to poor water quality (high ammonia or nitrites), we immediately perform a Hydrogen Peroxide Treatment. If after two treatments the problem persists, we perform a Salt Bath Treatment.

IMPORTANT

At the first sign of any problem, whether it is potential disease or water quality (high ammonia, low oxygen, etc.), immediately cease feeding all fish in the system until the problem is identified and resolved.

Hydrogen Peroxide Treatment

Hydrogen peroxide (H_2O_2) is commonly used in freshwater aquaculture for the treatment of external fungal, bacterial and parasitic infections. H_2O_2 naturally degrades to water and oxygen by a variety of chemical and biological mechanisms, including enzymatic decomposition by algae and heterotrophic bacteria. Bacteria account for the majority of degradation.

Due to its rapid degradation, hydrogen peroxide can be added directly to the culture tank without first moving fish to a quarantine tank or discarding the treatment water. Although hydrogen peroxide will not harm the biofilter at the recommended concentration, it can harm the fish at slightly higher than the recommended concentration by scarring the gill tissues and causing suffocation. Care must be taken to calculate the dosage correctly.

H_2O_2 Treatments last for a maximum of 60 minutes at a concentration of 50–75 ppm. Consecutive treatments may be carried out every second day to a maximum of three treatments.

To calculate the amount of H_2O_2 required for a concentration of 50 ppm (50 mg/L):

1. Calculate the volume of water in the tank by measuring the tank radius and the height of the water (using the formula for a cylinder):

$$(\pi) \times (Radius^2) \times (Height) = Volume$$

For example, in an 8′ tank with 3′ of water:

$$(3.14) \times (122 \text{ cm} \times 122 \text{ cm}) \times (90 \text{ cm}) = 4,206,218 \text{ cm}^3 = 4,206 \text{ L}$$

2. Calculate the correction factor for the strength of hydrogen peroxide you are using.

A correction factor is used whenever the additive component is not 100% active ingredient. In this case, the hydrogen peroxide product contains 30% H_2O_2 (the active ingredient) and 70% water.

For example, using 30% "technical grade" H_2O_2:

$$\text{Correction factor} = (100) / (30) = 3.3$$

3. Calculate the amount of H_2O_2 to be added to the tank:

$$(\text{volume of tank}) \times (\text{dosage of } H_2O_2) \times (\text{correction factor}) = \text{amount } H_2O_2$$

For example:

$$(4,206 \text{ L}) \times (50 \text{ mg/L}) \times (3.3) = 693,990 \text{ mg } H_2O_2$$

4. As H_2O_2 is a liquid, it is desirable to convert milligrams (mg) to milliliters (mL):

$$(693,990 \text{ mg}) / (1,000 \text{ mg per gram}) = 694 \text{ grams (rounded).}$$

$$(694 \text{ grams}) \times (1 \text{ gram per milliliter}) = 694 \text{ mL}$$

From this example, we can see that for a tank with 4,206 L of water, 694 mL of 30% strength hydrogen peroxide yields a concentration of 50 ppm.

To perform the H_2O_2 Treatment:

1. Isolate a tank by shutting off the water inlet and turning on the backup oxygen to that tank. Place a submersible pump in the tank to keep the water circulating and use the DO meter to monitor oxygen levels.
2. Measure the required H_2O_2, following the example above, and dilute it in a 5-gallon bucket of water.
3. Slowly pour the bucket into the tank over a period of at least five minutes, ensuring it is well mixed so that no fish are exposed to high concentrations

of H_2O_2. Monitor the DO levels and observe the fish for signs of stress, such as irregular swimming (on their side) or gasping at the surface. If stress is observed, double-check that you calculated the dose correctly and turn the water inlet on to dilute the H_2O_2.

4. After 45–60 minutes, turn the tank inlet back on. Remove the circulation pump and turn the backup oxygen off.

Salt Bath Treatment

A Salt Bath Treatment is effective in treating many common bacterial pathogens. The treatment causes a pathogen to enter a state of osmotic imbalance which kills it.

IMPORTANT

Salt is toxic to the plants and biofilter. It is critical that the treatment water is never allowed to enter the system water. Treatments are always done in the Purge Tank, and the treatment water is discarded afterwards.

Fish can be treated for 2–3 hours or more at a salt concentration of 1% (10g per liter) so long as they show no signs of irregular swimming or surface gasping. A concentration of up to 3% (30g per liter) can be used for short baths of 10–30 minutes so long as fish show no signs of stress.

1. Cease feeding the cohort 24 hours prior to treatment.
2. Isolate the tank from the system: shut off the water inlet and turn on the backup oxygen to that tank. Use the oxygen meter to monitor DO levels during the transfer process.
3. Use a submersible pump to transfer the water. Fill the Purge Tank ¾ full with water from the fish tank that is to be treated.
4. Turn on the oxygen to the Purge Tank and monitor DO levels during the treatment.
5. Calculate the volume of water in the Purge Tank using the formula described in the H_2O_2 treatment: $(\pi) \times (R^2) \times (H)$.
6. Add 10g of salt per liter of water in the tank, to achieve a 1% salt concentration. Dilute salt in a 5-gallon bucket of water before adding to Purge Tank over a 5-minute period. Use the submersible pump to circulate the water in the Purge Tank.

IMPORTANT

Use only pure sodium chloride (NaCl). Never use de-icing salt or anything with anti-caking agents or any other additives. Always check the MSDS of the product to confirm that it is pure. Many brands of pool salt contain additives. We use Windsor pool salt, widely available from pool supply stores.

7. Transfer fish into the salt bath as quickly and gently as possible. In order to catch all the fish, you will need to drain most of the water from the fish tank by siphoning or pumping it down to the sump or cistern.

   Leave fish in the salt bath for 2–3 hours, observing frequently for signs of stress, such as gasping at the surface or swimming on their sides. If any of these signs are noticed, double-check that you calculated the amount of salt correctly and dilute with more water until the fish can be transferred back to their tank.

   During the treatment, use the opportunity to clean and sterilize the fish tank. Scrub any algae and biofilm off the walls and floor and flush it down the tank drain. Sterilize the tank with a strong solution of hydrogen peroxide: 15–20 ml of 30% H_2O_2 per liter of water in a clean bucket. Scrub the entire tank with the H_2O_2 solution and let it sit for an hour before rinsing down the tank drain.

   To refill the tank, turn on the water inlet to reconnect it to the system. When filling the tank, monitor the water level in the sump. Top up the sump with cistern water as needed. Do not fill the tank directly with cistern water that has not been UV treated. The tank may be topped up with source water if you have no cistern.

IMPORTANT

Discard the salt water. Do not drain it into the system or Waste Tanks.

Once the tank is back online and fish have been treated in the Purge Tank for 2–3 hours, remove the circulation pump from the Purge Tank and transfer the fish back to their tank. Turn off the backup oxygen to both tanks and reset the oxygen system so that it is ready for the next emergency event.

Lab Analysis

To be certain of the disease or pathogen infecting your fish, you will need to send one to a lab for analysis. The lab will want the fish as fresh as possible, ideally live, which will require shipping the fish on ice. Early on we sent fish to a lab on several occasions. We have not had to do so for a few years, as we have found that close fish monitoring, H_2O_2 Treatments, Salt Bath Treatments and basic biosecurity protocols resolve all issues.

Purging Before Harvest

Purging is required for fish in all recirculating aquaculture systems, especially in aquaponics. Recirculating systems provide an excellent environment

Purge tank with submersible pump for aeration and circulation.

for the growth of problematic varieties of bacteria, notably actinomycetes and cyanobacteria, and some species of algae which produce two metabolic waste compounds: geosmin and 2-MIB. These organic molecules are readily absorbed by fish and can impart a strong muddy or earthy flavor to the flesh. They are detectable by human palates at concentrations in the parts-per-billion range.

Purging, sometimes known as depuration, is a process of eliminating the bacteria and algae from the water and allowing the geosmin and 2-MIB to diffuse out of the flesh. This is done by holding the fish without feed in the Purge Tank for 10–15 days and exchanging at least 25% of the purge water with clean source water every day (50% for the first two days).

As the Purge Tank is isolated from the rest of the system, oxygen must be supplemented via the backup oxygen system or by using a submersible pump in the tank to lift water and spray it back into the tank. Withholding feed 24 hours before transferring fish to the Purge Tank will dramatically decrease their oxygen consumption.

Ammonia buildup is not an issue because of the lack of feed and daily water changes. Temperature should not be an issue as long as the Purge Tank does not go above 20°C. If temperature is an issue, use a circulation pump connected to a water chiller. Temperatures below 10°C are preferred as they will inhibit the growth of the undesirable bacteria and decrease fish metabolism, which in turn will decrease oxygen consumption and ammonia production.

A small UV sterilizer connected to the circulation pump is a beneficial addition to this process, as it will destroy any undesirable algae or bacteria that may be present in the purge water. The UV sterilizer will not destroy the geosmin or 2-MIB compounds produced by these organisms, so daily water exchanges must still be performed.

Do not harvest a cohort all at once, as this will remove too much biomass from the system. The number of fish you harvest will be dictated by total biomass of the system and your market demand.

For average 1 kg fish, we recommend filling the Purge Tank half full to purge 20 fish. Fill the Purge Tank to three-quarters full for up to 40 fish. You may be able to purge more biomass than this in a full Purge Tank, but we recommend harvesting no more than 20–30 fish at a time.

Purge Procedure

1. Sterilize the Purge Tank with hydrogen peroxide: add 15–20 mL of 30% H_2O_2 per liter of source water in a clean bucket. Scrub the entire tank. Let sit for an hour then rinse down the tank drain.

2. Fill the Purge Tank with an appropriate amount of water for the biomass to be purged. Use 50% system water and 50% source water. The temperature must be within 5°C and the pH within 1.0 of the water the fish are being moved from. Use more or less source water if needed to bring the Purge Tank within these parameters.

3. Place a submersible circulation pump in the tank or turn on the backup oxygen to the Purge Tank. Monitor oxygen level with the DO meter.

4. Move fish to the Purge Tank. Try to move the largest fish, although they are typically faster and harder to catch.

IMPORTANT

Do not feed fish being purged at any point in the purge process. Additionally, cease feeding the whole cohort for 24 hours prior to moving fish to the Purge Tank. Resume feeding the cohort once fish to be purged are moved.

5. For the first two days, exchange the purge water by pumping or siphoning 50% of it into the sump and refilling with source water.

6. After two days, only 25% of the purge water needs to be exchanged daily for the duration of the process. Never use cistern water or system water to purge, as it contains the undesirable bacteria and organic compounds. Use only source water for daily replacement. Water used in the purge process can be added back into the system or stored in the cistern.

7. After a minimum of ten days, test the quality of your fish by cooking and tasting a fillet without seasoning or butter. If you taste even a hint of muddy or earthy flavors, continue purging. Always taste test your fish before selling them, as conditions are inherently variable and the time it takes for the organic compounds to diffuse from the flesh can vary with conditions.

Harvesting Fish

Let's be honest at the outset: when we talk about harvesting fish, we are talking about killing. There are many methods and ethics regarding the harvesting of fish. We have tried many methods, including CO_2 gassing and sedation. In our experience, and according to studies done by the Humane Society of the

United States, the quickest and most assured method of killing a fish, and thus the most ethical, is also the simplest and oldest: hitting it on the head with a blunt object. We use a hardwood fish bat for this purpose. It has the right weight and fits the hand nicely. You will likely need to experiment to find the object that works best for you.

The method is very simple: remove a fish from the Purge Tank and while it is still wrapped in the net and easy to handle, strike it hard on the head a few times. A good hit will instantly kill it. With a little practice, this method is the quickest, surest and most humane. Remove the fish from the net and cut its gills to bleed it.

Optionally, just before harvest, you can put bags of ice into the Purge Tank to quickly lower the temperature, putting the fish into a nearly catatonic state. Doing this will also prechill the flesh and delay rigor mortis, which may or may not be desirable depending on your market.

Freshly killed fish with fish bat and knife.

Confirm with your local health authority what level of processing you are allowed to do onsite. Depending on your intended market, you may require certification for processing. In some cases, you may be able to remove the guts and gills before selling as whole fish, direct to consumer, without requiring a slaughter or processing permit.

If you want to further process fish onsite, you will likely require a specific food-processing license that may require the addition of a separate health authority-approved building or kitchen. Always check your local regulations.

Harvested fish must be immediately put on ice, moved to the cooler or frozen.

Plant Production

PLANTS ARE THE PRIMARY SOURCE OF INCOME on an aquaponic farm. Our multistage production system is designed to maximize space at all stages of growth, thus maximizing yields and profit.

The DWC troughs are the most valuable real estate in your farm. The priority is to minimize the time required by each plant to come to maturity and to have a constant supply of new seedlings ready to be transplanted into the troughs. The biggest challenge, other than control of the environment, is applying a human timeline (a weekly planting schedule) to an ecosystem that grows at its own pace and changes with the weather.

Plant Selection

Many plants thrive in a cold-water DWC system, and many do not. The key is to find plants that thrive in your system that you can easily market at a good price. The parameters of the system are not as adaptable as your choice of plants. If a plant does not thrive or you cannot easily sell it, grow a different plant.

In general, you can separate plants into two categories: those with slower growth that will be grown in stages and those with faster growth (6 weeks maximum from seed to harvest) that will be directly seeded into the troughs.

The Plants We Produce at RCA

We have tried the following plants/families and found that they thrive in our system: lettuce, mustards, watercress, kale, mint, the choi varieties, cilantro, Swiss chard, leeks, beans/peas and celery.

We have tried the following plants and found they do poorly in our system: spinach, strawberries, tomatoes and basil. Spinach in particular was surprising to us as it is a cool/wet season crop, but it seems that it does not like wet feet (roots constantly wet).

There are many crops we haven't tried but assume to be ineligible for other reasons: root crops such as carrots and potatoes may rot underwater, winter brassicas like cabbage and broccoli simply take too much space and/or too long to grow, and other crops like corn and squash are physically incompatible due to their size and/or spread.

Even though they grow well, we do not produce mint, the choi family or Swiss chard, as they are challenging to market or the market price is too low. We do not produce brassicas or beans/peas, as they take too much space and time, nor cilantro which takes a long time to sprout and is prone to bolting in our climate.

Some plants, such as watercress, grow exceptionally well. We produced large quantities only to find that while it is in demand and sells at a premium, the local demand is much less than our production capacity, and we ended up with considerable unsold stock that became compost. Determining the right balance between a plant's capacity to thrive in your system and the local demand for it will take experimentation.

Of the plants that thrive, we have narrowed down what we produce on a regular basis to three varieties of lettuce, three varieties of mustards, watercress and baby kale. We also grow small amounts of leeks and celery on a seasonal basis. For us, these plants have the best balance of speed of production, ease of marketing and the highest profits from sale.

We grow green leaf, romaine and butterhead lettuce. We recommend experimenting to find the types that thrive in your system, then reducing your regular production to 3–6 varieties. We grow Tropicana green leaf which is large and robust, Coastal Star romaine which has a classic blanched heart structure, and Alkindus and Roxy butterheads which are a brilliant red color with tender blanched hearts. We have tried dozens of other varieties of lettuce, with varied results, and continue to test new varieties regularly.

Mustards grow very well in our system. We grow Giant Red, Mizuna and Red Kyona Mizuna. All three thrive and produce vibrant, thick leaves with varying levels of spice (Giant Red is the hottest).

Watercress is perhaps the most vigorous plant we've ever grown. We grow true watercress (*Nasturtium officinale*). In contrast to curly cress or peppercress, true watercress is a rich, dark green with robust stalks and leaves. It is firm and crunchy with a spicy fresh flavor, grows quickly and is very resistant to pests and diseases.

We grow a variety of kale called Abundance, typically as a baby leaf, though it can be grown to full size. We harvest ours young to add to our salad mix or sell in bulk.

Both watercress and kale are fast-growing plants that can be directly sown into the troughs (no separate sprouting or seedling stage) to save on labor. Because of the time it takes to grow them to maturity, lettuce and mustards are sprouted in the Germination Chamber and grow in the seedling area prior to planting in the troughs.

On average, lettuce accounts for approximately 75% of the plants we produce. We produce a little less romaine than green leaf and butterhead. The remaining 25% is split between watercress, mustards and baby kale. The exact planting of each crop depends on a variety of factors, notably what grows particularly well during that season, which seedlings are the strongest and what is in most demand at market.

Seeds

There are numerous seed producers to choose from. Sampling and experimentation will be required to determine the best producer for your farm. In general, we recommend producers of organic, non-GMO seeds that are located near you and/or within a similar plant hardiness zone. We buy all of our seeds from West Coast Seeds, a BC company. We recommend them.

The quality of seeds will vary from year to year, even from the same producer. Raising plants (and seeds) is always impacted by nature. Good seed producers will be able to somewhat buffer the effects of an off season, but seed quality variance should be expected, particularly from producers of organic, non-GMO seed.

For lettuce, look for varieties that grow vertically more than horizontally. This will allow more plants per raft without crowding, and low-growing (horizontal) lettuce tend to suffer more from bugs and rot around the stems. Romaine is an excellent example of a lettuce that grows vertically. Butterhead is an example of a lower, wider-spreading plant.

IMPORTANT

If available, we strongly recommend pelleted seeds. Pelleted seeds are encased in a small ball of clay which quickly dissolves when moistened. They are substantially easier to handle, making it quicker to plant one seed per cell, which eliminates thinning later.

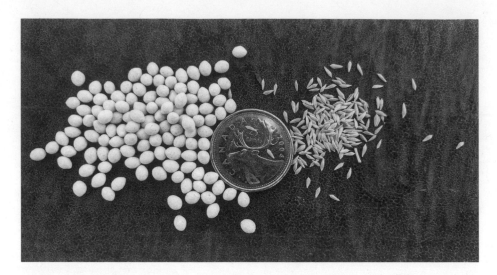

Pelleted (left) and non-pelleted (right) lettuce seeds shown with a quarter for scale.

Seeds must be stored in a cool, dry, dark place that is protected from bugs and vermin. Improper storage can spoil or greatly reduce their viability.

The Growth Cycle

The growth cycle of a plant will vary depending on the time of the year, primarily due to fluctuations in photoperiod and air temperature. The two types of production are multistage (transplanted) and direct seeding.

A typical multistage growth cycle is:

- 1 week sprouting in the Germination Chamber.
- 3–4 weeks growing first leaves on the Seedling Table.
- 3–6 weeks growing to maturity in the troughs.
- Note that larger plants are thinned or spaced half way through the trough stage.

Lettuce, mustards, cilantro, Swiss chard, leeks, beans, peas and celery are all grown in a multistage cycle as they all take four weeks or more to grow into a 3–4″ seedling from seed. Although a multistage cycle makes more efficient use of space, it also involves more labor due to the additional steps of transplanting and spacing/thinning.

Direct Seeding

Direct seeding is sowing seeds directly into net pots and inserting into troughs where they will germinate and grow until harvest. Watercress and baby kale are grown in this manner because they can reach a harvestable size in 4–6 weeks. The goal is a dense canopy that can be harvested en masse. Spacing/thinning is not required.

For each raft:

1. Place a thin layer of transplanting media (coarse coco chips or loose rock-wool) at the bottom of each pot. The purpose of the transplanting media is to create a physical barrier that prevents the growing media around the seedling from washing out. A few flakes of coco chips or ¼″ of loose rock-wool will suffice.
2. Add a layer of germination media (coco coir or Pro-Mix) on top of the transplanting media. There is no need to presoak the media as it will wick up enough water from the trough to moisten the seeds.
3. At the workbench, insert the pots into a raft and sprinkle seeds directly on top of the media. Baby kale is seeded at 10–15 seeds per pot; watercress at 15–20 seeds per pot.
4. Place raft in trough. If the air is very dry, it may be necessary to loosely cover the raft with a sheet of clear plastic until the seeds germinate. The media will wick up the water from the trough, moistening the seed and starting germination. Both watercress and kale germinate in 2–3 days under ideal conditions.

If you used a plastic covering, remove it once the seeds have sprouted. Once the crop has begun to form a canopy, rafts should not be moved until harvest. There is no need to thin or space baby kale or watercress.

Multistage Production

Planting Seeds

The Germination Chamber holds 18 seed trays, each with 98 cells. This means you will be seeding 1,762 new cells every week. Like all aquaponic tasks, efficiency is key, so think of it like a factory line.

To speed up the process of pressing holes into the media for the seeds (dibbling), use a dibbler plate that can dibble a whole tray at a time. See Dibbler Plate, Chapter 5.

1. Fill the workbench with trays, bumper to bumper.
2. Spread seedling media (coco coir or Pro-Mix) over all trays, gently filling the cells.
3. Use a submersible pump, hose and diffusing wand to water the trays with trough water (use the same setup to water seedlings).
4. The water will cause the media in the cells to compress. Gently add additional media to fill all cells. No additional water is required, as the additional media will wick water up from the wet media.
5. Use a dibbler plate to press shallow holes into each cell.

6. Place one lettuce seed in each cell, or 1–3 if using non-pelleted seeds. If you are growing plants that are intended to grow as a bunch but are slower growing thus not direct seeded (e.g., cilantro), place 5–10 seeds in each cell. For mustards, place 3–6 seeds per cell.

We have tried numerous tools and gadgets to make the seed planting process more efficient. Nothing has proven any faster or easier than seeding by skilled hand. Pelleted seeds will greatly improve efficiency as they make it very easy to place only one seed in each cell. If using non-pelleted lettuce seeds, you will need to practice placing a small "pinch" of 1–3 seeds in each cell, and these will need to be thinned to one per cell after germination. Buy pelleted seeds if possible.

Unless they must be germinated in darkness, do not cover the seeds with the growing substrate, simply leave them on the surface. Do not add more water. Place a low-humidity dome over each tray to keep seeds from drying out and move them to the Germination Chamber.

Newly seeded trays with dome lids.

IMPORTANT

Many seeds can be germinated directly on top of the media (not covered by media) as long as the tray is covered with a humidity lid, while other seeds must be sown under the media. Lettuce, for example, will not germinate without some light on the seed, whereas cilantro and parsley will only germinate in complete darkness. Refer to the instructions provided by your seed manufacturer.

Germination

Seeds will spend one week in the Germination Chamber. This is not a flexible timeline, as it is necessary to have a constant supply of seedlings ready to transplant, in order to operate at full capacity. If you find that plants do not sufficiently sprout in one week with good environmental conditions, consider different varieties.

The Germination Chamber is kept at 17–18°C. With the recommended heating and cooling system (see Chapter 4), it should be possible to keep the chamber consistently in this range in even the most extreme weather. Each of the two chamber tables holds nine trays, with each table lit by three 3′ T5 bulbs. We recommend 20 hours of light per day.

During the week of germination, your only job is to ensure the correct temperature in the chamber, which is automatically controlled by the thermostat. Do not open the domes and do not add any water.

It is possible during the milder seasons to germinate in the greenhouse on the Seedling Table, but we recommend against this. Your Germination Chamber is designed to do a specific job and once dialed in should germinate a high percentage of seeds rapidly. Additionally, when you are operating at capacity, all production areas should be full (or very close to full) at all times so you should not have room on your Seedling Table.

Seedlings

After one week, the trays are moved from the Germination Chamber to the Seedling Table to make room for the next week's planting. Place newly germinated trays directly under the grow lights as younger plants (once sprouted) require the most light. This may require a bit of strategic rearrangement.

Leave the domes on the trays for 2–3 more days unless the sprouts are so tall that they begin to touch the dome.

For most of the year, seedlings are watered once a day, twice if it is particularly dry. Water only in the morning or the evening — never in the middle of the day or in very intense sunlight. In the winter, high humidity makes it harder for water to evaporate from leaf surfaces, and standing water left on the leaves from irrigation can smother them and invite disease. During these times, it is better to water the seedlings less (every other day, as long as they aren't drying out), and to water only in the late morning so that the leaves have a chance to dry during the day.

If you sowed more than one seed per pot for plants that will grow one plant per pot (such as non-pelleted lettuce seeds), trays will need to be thinned before the cells get too crowded (typically 2 weeks after germination). Gently

Our Seedling Table.

pull out the smaller or weaker seedlings, leaving one healthy seedling in each cell.

Seedlings should grow on average to 2–3″ tall in 3–4 weeks in the seedling area. The conditions of the seedling area will be the same as the rest of the greenhouse. The supplemental HID lights are used on the same schedule as the supplemental trough lighting. The lighting schedule is based on the target photoperiod (~14–16 hours) and the amount of supplemental light needed for the time of year. See Supplemental Lighting in Chapter 2.

Transplanting into the Troughs

The primary factors for determining which seedlings to transplant are the maturity of the seedlings and your knowledge of what your market demands. Once you have an established market, plant maturity is typically the determining factor.

The number of rafts to transplant will vary throughout the year primarily due to the length of time plants take to grow to full maturity in the troughs. Our lettuce is ready in as little as three weeks under perfect conditions and can take as long as six weeks in the winter. The quantity of seeds planted can also be adjusted once you are familiar with the seasonal effects on your production timelines.

Our high-tech custom transplanting tool.

If you achieve a 95% success rate for germination and seedling growth, each seedling tray with 98 cells will fill approximately three rafts (96 sites). As you should not expect to achieve a 95% success rate until you have had some practice, both the Germination Chamber and Seedling Table are oversized.

You need the following items for transplanting: washed rafts, clean net pots, transplanting media and a transplant tool. All items should be located next to the workbench for efficiency.

Unless the seedlings are large and root-bound, you will need a transplant tool to remove them from the cells and place them in the pots without damage. Make your own custom tool by taking a kitchen fork and bending the tines to create a small spade. With some experimentation, you can create a shape that feels good in your hand and fits perfectly into the cells. You will be doing this action many thousands of times — extra time spent creating the best tool will pay itself back many times over.

1. Place 32 net pots on the workbench.
2. Place a thin layer of transplanting media (course coco chips or loose rockwool) at the bottom of each pot.
3. Gently remove a seedling with the transplant tool and place in the net pot. Gently seat the plug in the coarse media.
4. Repeat to fill all pots. As much as possible, plant each raft with the same variety.
5. Place a raft on the workbench and fill with transplanted net pots then place the raft into a trough.

Repeat until the troughs are full or according to your transplant schedule.

Transplant media in bottom of net pot.

Gently remove remove seedling from its cell with transplant tool.

Place seedling into net pot.

Left: *Freshly transplanted seedlings.*
Right: *Place net pots into raft. Place loaded raft into trough.*

Raft Placement and Rotation

The placement of rafts should be done in a conveyor method. Water enters the west end of the north side of each trough, flows to the eastern end then loops back, flowing east to west down the south side of the trough. The highest concentration of nutrients is found at the trough inlet, just after the fish tanks which produce the constituent nutrients and the CFB where they are converted into minerals. The younger a plant, the more light and nutrients it requires (think how much more a teenager eats than an elderly human).

A plant (and human) will always grow bigger and healthier if they are given ample sunlight and nutrients when they are young, even if the light and nutrient content diminishes when they are mature. Accordingly, the newly planted rafts are placed in the most upstream spaces in the troughs. As the plants grow, the rafts are moved downstream. Each week the most mature rafts are harvested from the outlet end of the troughs, and all the remaining rafts are shifted downstream to make room for new seedlings at the inlet end. During this process, any crops that have begun to crowd should be spaced or thinned to give them more room.

Transplant Schedule

For a number of potential reasons, it is unlikely you will be able to keep an exact transplanting schedule. Most commonly, this is due to the varying seasonal production times. It is important to always think ahead to your market.

For example, if you know (or estimate) that your production time is five weeks, you should be harvesting and transplanting ⅕ of your rafts weekly (approximately 50 rafts). If for some reason you have ⅓ of the trough space available for new plants, perhaps due to disease or a one-time large-volume sale, strive to fill all the rafts through a combination of transplanting seedlings and double-spacing semi-mature rafts (rather than thinning). As much as possible, use all growing space at all times. Think ahead and make a plan for selling the excess production, such as a sale price on certain products at your market booth or wholesaling the excess.

The goal in an aquaponic farm is a consistent, year-round supply at the maximum efficiency of the system with maximum efficiency of labor. This means keeping the troughs fully planted at all times. Once you are familiar with your system, including seasonal production changes, and know the demands of your market, you will be able to create a workable transplant schedule, but it is important to remain flexible as farming requires adaptability.

IMPORTANT

Remember that "production time" in this book means the time a plant spends in the troughs. Germination and seedling stages are never included in "production time."

Thinning and Spacing Plants

Our rafts are designed to hold up to 32 plants at 6″ centers. With many plants, notably lettuce, this is not sufficient spacing to produce to maturity. Typically around half-way through their time in the troughs, plants will start crowding each other, causing numerous problems. Notably, they will deform as they stretch to compete for light, and having a solid canopy of plants allows bugs and disease to flourish in the moist dark areas underneath the canopy and around the stems.

We grow plants on 6″ centers for roughly the first half of their production time as this is the best use of space (our primary hydroponic design principle). Sprouts fit into less than 40 square feet in the Germination Chamber;

seedlings fit into 125 square feet on the Seedling Table, then they are spaced at 32 per raft until they start crowding. Once the leaves begin to touch their neighbors, the crop is spaced to 16 plants per raft by thinning or spacing to yield a plant in every other hole, which will look like diagonal rows of plants. At 16 sites per raft, spaced diagonally, plants will be on 8.5″ (diagonal) centers — sufficient for most plants, including lettuce, to grow to maturity.

Thinning means removing plants altogether from the troughs. If the plants are particularly weak or diseased, they are composted. If the plants are simply stunted or otherwise less vigorous, the acceptable parts can be used in a salad mix. When thinning a raft, remove half the plants by choosing the smallest and weakest or any that are showing signs of pest or disease, and then gently rearrange the remaining plants into the double-spaced configuration.

Spacing means moving every other plant to an empty raft. After spacing, the 32 plants from one raft will occupy two rafts (16 per raft).

IMPORTANT

Whether thinning or spacing, use the opportunity to quickly inspect for and remove any moldy leaves from the undersides of the plants or the top of the rafts.

The decision of whether to thin or space plants will be based on the quality of the plants and the needs of your market. If, after harvesting, there are only enough empty rafts to hold this week's transplants, thinning the crop is more appropriate. If there are extra empty rafts, perhaps due to a large harvest or a previous lack of transplantable seedlings, spacing the crop is more appropriate. On most occasions, you will be both thinning and spacing. Thinning and spacing takes place after harvest cleanup and before transplanting, thus allowing you to gauge how much free space there is for the new transplants.

IMPORTANT

After thinning or spacing, place a clean empty net pot into the open holes of every raft to shade the water from excessive sunlight, which can harm the biofilm on the bottom of the trough.

Butterhead lettuce: single spacing (right raft) and double spacing (left raft).

Maturing Plants

After a crop has been thinned and/or spaced, the plants will soon begin to touch again and will eventually form a canopy. This is acceptable as the process of spacing has created enough airflow around the base of the plants to lower the risk of mold and sufficient room to prevent competition for light.

Once the plants form a canopy, it is best not to jostle the rafts or pull them out to move them, particularly in the case of lettuce. Simply float them downstream when they need to be rotated through the trough, according to the conveyor method. As the plants form a dense canopy, they become interdependent for the physical support needed to stay upright. If a mature crop is thinned too late or jostled vigorously, they will fall over and have to spend their energy over the next few days turning back to an upright position.

Plant Inspection

There is very little ongoing work required for the plants themselves during their production time other than inspection and correcting problems. The primary role of an aquaponic farmer is to ensure that the ecosystem is functioning within the parameters specified in Chapter 6. Fluctuations in pH, water temperature, ammonia and nitrites must be remedied immediately.

Plant inspection takes place via a quick daily check and a more thorough weekly evaluation. Every day, walk down the aisles and observe the rafts and Seedling Table. Notice any symptoms of disease or nutrient deficiency. Notice leaf discoloration or wilting. Notice obvious signs of pests or swarms of pests. Notice any mold or mildew. Mostly notice any changes from the previous day. If you monitor daily, you will learn to spot problems early.

Weekly, do the same but much more closely. Whereas your daily inspection may take less than five minutes, your weekly inspection may take an hour. Scan the rafts and seedling trays closely and briefly look at each plant individually. Focus in on any symptoms. If you find a symptom, look more closely at the other plants in that crop and particularly at the other plants of the same variety. Is the symptom localized to one plant, or in one area, or to one variety, or is it throughout the crop or the whole trough? Check sticky traps for signs of pests and focus on plants in those areas.

See Nutrient Deficiencies later this chapter and Pest Management in Chapter 10.

IMPORTANT

A quick daily inspection and thorough weekly inspection are your best tools for identifying nutrient deficiencies and preventing the spread of pests and diseases. With practice you will be able to quickly identify and correct potential issues before they spread and become a serious problem.

Watering in the Greenhouse

The greenhouse environment is not far removed from nature, with natural sunlight and free air movement, but it lacks one fundamental force of nature: rain. Rain provides more benefits to crops than just water. It washes pests and dust from leaves. It can drown pests. Powdery mildew favors humid conditions but only grows on dry leaves, and its spores are smothered by water temporarily pooling on leaves. Plants also absorb some waterborne nutrients through leaves, particularly calcium, potassium and iron.

After considerable experimentation, we found that watering plants with water from the troughs is as effective as botanical and biological pesticides at keeping most pest and disease problems in check. Watering is not a replacement for pesticide use, which still is required on occasion, but it works very well as a prophylactic maintenance program. Using this practice, we have cut our use of pesticides by as much as 75% and only require them when an outbreak is not being effectively suppressed by a combination of watering, cultural and environmental controls.

Watering with system water also provides nutrients directly to the leaf surfaces and helps to establish a healthy microbial ecology in the crop. A strong microbiome is the very first line of defense against both aquatic and terrestrial pathogens. A healthy population of beneficial bacteria will outcompete pathogens and prevent them from establishing in sufficient numbers to cause disease. A similar process is at work in your digestive tract and the biofilter of your system.

If you water your plants with system water, we recommend rinsing produce with potable water before selling.

In consistently hot and dry seasons, we recommend watering all plants twice per week, in the morning or evening when the sun is less intense. In the winter, when humidity is high and water is slower to evaporate, reduce watering to once per week or less and only in the morning so the water can evaporate during the day. It is important to avoid having standing water on leaf surfaces for extended periods of time as this can promote mold growth rather than prevent it and it can suffocate the leaves.

Do not water the plants within three days of spraying a pesticide, in order to allow the pesticide to work. Also, if the pesticide is potentially harmful to the fish, the time will allow it to naturally degrade in the air and sunshine. See Chapter 10 for more information on pesticides.

Making it rain in the greenhouse.

To water the plants, place a small (~⅓ horsepower) oil-free submersible pump in the water near the outlet end of one of the troughs. Attach a garden hose and a watering wand with a diffuser to form a gentle spray. Use this same pump and nozzle for watering the seedlings.

One of the added benefits of growing in DWC troughs is that any excess water sprayed onto the plants will drain back into the troughs with only a small amount lost to evaporation. In a tower system, watering plants in this manner will waste large quantities of system water.

Harvest, Packaging and Storage

Harvesting Methods

The decision to harvest is based on a combination of your market and plant maturity. Leafy greens (e.g., lettuce) that are left growing past maturity will start to bolt (transition to producing seed). At this stage, the quality will progressively deteriorate as the plant quickly stretches upward and becomes very bitter. Conversely, you may opt to harvest some plants prematurely to fulfill market demand. If harvesting early, it may be appropriate to adjust the sales price accordingly. The decision of when to harvest is something you will learn to adapt to as you get to know your farm and your market through the seasons.

The harvesting process varies depending on where/how you sell the product and whether plants need to be washed. We find that some restaurants want dry, never washed head lettuce while others only want washed and bagged mixes. If you are selling in bulk or wholesale, individual bagging is not usually required, but you may find it preferable when selling at a farmers market.

Most harvesting occurs at the troughs. We find this cleaner and more efficient than carrying laden rafts to the workbench. If you are producing the types of plants we do for the same end products (heads of lettuce, salad mix, herbs and other preweighed produce), you will need at least four bins (totes) with you at the trough: one for each variety of strong lettuce to be sold as heads, one for "inglorious" plants to be processed into salad mixes, one for compost and one for net pots. We separate our bins into clean bins that are only used for produce and dirty or older beat-up bins that are only used for compost and net pots.

Other plants (on our farm, mustards and baby kale) that are harvested in a cut-and-come-again style may be preferable to harvest by taking the entire raft to the workbench, where it is easier to pick off individual leaves without removing the plant from the raft. Use separate totes for compost and net pots. You will need several totes for products to be sold.

Depending on what is comfortable for you, you may harvest some or all of the pots while the raft is in the trough or you may pull the raft partially and/or fully out of the trough as you harvest. Experimentation is required to find what works best for you.

To harvest a lettuce or other head-type plant, pull the whole plant — plant, net pot and roots — out of the raft. Twist off the roots that hang through the pot and put into the compost bin. Separate the pot from the root ball. Put the pot in the net pot bin. Cut the root ball off the plant and put it in the compost bin. By hand, remove all damaged or ugly outer leaves by pulling the leaf out and down towards the base of the plant. Put all discard leaves in the compost bin. Put lettuce into a harvest bin, separating plants to be sold as heads and those for mixes.

In summer, head lettuce displayed at a farmers market will dry out and wilt. This is unappealing to customers and will result in fewer sales. To keep heads fresh, they can be harvested and bagged fully intact, roots and all, and displayed live at the market in shallow bins with a bit of water in the bottom.

For live sales, remove the lettuce from the raft, leaving roots and net pot intact. Remove damaged or ugly leaves then put into a tote with ½″ of water in the bottom, roots down. At the market, use shorter bins (with water in the bottom) to display the crisp heads. When sold, leave the root ball attached or cut off depending on customer preference. Displaying lettuce like this is a big draw and is an excellent icebreaker for starting a conversation about the benefits of aquaponic farming. In addition, unsold lettuce can often be placed back into the rafts to continue growing, as long as care is taken to prevent damage during handling and transport.

We often sell butterheads as living lettuce, which keep for a very long time, even unrefrigerated. This is a value-added product which is often in high demand and sells for a premium. To harvest a living lettuce, twist and remove the hanging roots and net pot but do not cut off the root ball. Use the bottom of a small plastic bag and an elastic band to contain the moist root ball, then pull the bag up around the lettuce, leaving the rosette of leaves exposed.

Watercress, when seeded at the recommended rate, will produce a thick mat of individual shoots and leaves with a layer of roots spreading across the top of the rafts. We find that cress does not regrow well in a cut-and-come-again harvest, so it is more economical to fully harvest and reseed. To harvest cress, pull out a net pot and gently tug the shoots to disentangle them from the rest of the mass. Notice the small white roots growing from the stems just above the root ball. Leaving these roots on can greatly extend the shelf life (1–2 weeks vs 5–6 days). If the cress is being sold by itself, cut the root ball low enough to leave some of these roots on. Place bunches in a bin with shoots and roots all

Remove roots below the net pot.

Remove the net pot. Leave root ball intact.

Remove damaged leaves.

Wrap plastic bag around base of lettuce.

Use elastic band to secure bag around the root ball.

Freshly harvested living butter lettuce.

facing in the same direction as this will make them easier to handle for portioning and packaging. If the cress is destined for a salad mix, cut the bunch higher (above small white roots) leaving only shoots and leaves. See color photo.

Some plants can be harvested more than once using the cut-and-come-again method. We use this technique to harvest mustards and baby kale in two slightly different ways. When using this method, try to avoid taking more than 30% of the plant at any one time.

Mustards typically yield 2–6 harvests before growth slows, depending on the variety. They don't tend to produce a thick canopy of leaves, so it is possible to snip or cut individual leaves from a plant without removing the net pot from the raft. Do not remove the leaf by breaking it off by hand. Separate leaves into two harvest totes: larger leaves for individually packaged units and smaller leaves for salad mixes. If we have an excess of mustards, we offer them for sale separately, but we always prioritize having sufficient supply for mixes.

Baby kale grows a dense canopy of leaves when seeded at the recommended rate and will also regrow vigorously enough to harvest 2 or 3 times. It is "mowed." Without removing the individual net pots, cut all leaves off at a relatively uniform height, leaving around 2–4″ of plant to regrow. Try to cut at a height that will harvest most of the mature leaves, while leaving the growing tips at the center of each shoot. In 2–3 weeks, the plants will yield another harvest.

We sell romaine as full heads and hearts. Hearts are the innermost part of the plant. Depending on market demand, we typically sell larger undamaged plants as heads and smaller damaged and/or deformed "inglorious" plants, with outer leaves removed, as hearts. Hearts are sold in bags of 2 or 3 (approximately 1 lb total per bag). They are often in high demand, sell for a premium (though not as much as the same number of healthy large heads) and are an excellent use for inglorious plants that may otherwise be unsellable. Put heads and hearts in separate harvest totes.

IMPORTANT

Handle all plants as gently as possible. Stems and leaves break and bruise easily, and damage will become more and more apparent as the harvested plants age. When filling bins, do not overfill, as this will damage plants at the bottom.

Washing Plants

Some plants may need to be washed, depending on the market and the quality of the plant.

Most head lettuce we sell at a farmers market are unwashed unless they have more than a few insects on them or were sprayed within two days prior to harvest. Our salad mixes are always washed twice and dried before packaging.

One of our regular restaurant clients insists that lettuce remain unwashed to extend its shelf life. They prefer to wash it themselves immediately before use. Wet produce will have a shorter shelf life unless the produce has ample opportunity to dry before packaging.

Washing your produce will remove bugs and dirt and can also help to reinvigorate plants that have wilted. If you are not using a walk-in cooler, washing your produce in cold water will help to chill it and extend its shelf life prior to shipping.

IMPORTANT

If your produce is washed, only use fresh potable water. Never use system water to wash your produce before sale.

The easiest method of washing plants is to place the loaded harvest bins on the sump lid and fill them with potable water until all plants are submerged. Let plants soak for 5–10 minutes before draining. The water used for rinsing produce can be drained directly into the sump so long as there are no leaves or bits of plants. We pour the water through a piece of window screen as a filter.

If you are attempting to reinvigorate wilted head lettuce, the best way to do this is to make a fresh cut across the base of the plant, where the root ball was removed, and place the head in lukewarm water. This will allow the plant to absorb water far more quickly than using cold water. After 15–20 minutes, the heads should look fresh and crunchy again, ready for bins or packaging.

Salad Mixes

Salad mixes are always a big seller at a farmers market and often sell at a premium. Customers appreciate having a "ready to eat" option. For an aquaponic farmer, making salad mixes is a great way to sell inglorious plants as a premium product instead of composting them. We typically find about 25% of our head lettuce are inglorious (less than ideal in size or presentation but still perfectly edible). Always remember that even inglorious heads can still have good hearts.

In order to make a good mix, variety is required, not just lettuce. We grow varieties of mustards and watercress specifically for adding to our mixes. Our mix is atypical, as it is "assembled" from various plants that are grown individually. Most conventional salad mixes are grown by broadcast seeding a variety of seeds together in one bed before mowing the crop. This method is a very fast and easy way to produce a mix comprised entirely of baby leaf greens, but the farmer has no control over the amount of each component. Our style of salad mix is very well received, as it is always crunchy and robust with an excellent shelf life.

We try to make our salad mix with approximately 50% lettuce (three varieties) and the rest a varying mix of mustards, baby kale and watercress, depending on what is available that week. After harvesting plants that will go into the mix, rinse them in their totes with potable water (not system water). Submerge the plants for 5–10 minutes before draining, then store totes in the cooler until you are ready to assemble the mix.

To assemble the mix, remove the required totes from the cooler and stage the components on the stainless steel counter/sink. In one single cut, remove the base of each lettuce head and separate the leaves into clean bins. Discard any leaves that are too big or bug-eaten (this will mostly be done when harvesting). Assemble the components in clean bins in layers to minimize the need for mixing, which will cause bruises.

Once a bin is full with a mix, rinse a second time by filling the bin with potable water (not system water). Leave the mix submerged for 5–10 minutes before draining the water.

To ensure maximum shelf life, the mix must have as much water as possible removed prior to bagging. Use a commercial salad spinner. If you find you are processing large volumes (e.g., more than 100 lbs of mix every week), this may become an inefficient drying method. For more ideas on spinning and drying salad mix, we recommend consulting *The Urban Farmer* by Curtis Stone (New Society Publishers, 2015).

Once plants have been harvested and washed (if necessary), you are ready to portion and package. Always keep produce in the cooler after washing until you are ready to package it.

Portioning

When portioning your products, it is important to consider the price of each unit and your market. Certain products like salad mix and watercress are sold by weight, whereas others such as romaine hearts are sold by number. For some markets, it can be advantageous to offer a smaller and larger size of a product. With some practice, you will be able to determine which sizes and prices work best for each product and your customers.

We typically portion salad mix in 5 oz. and 8 oz. bags, mustards and watercress in 6 oz. bags and romaine hearts (3 per pack) in approximately 1 lb. bags.

Packaging

We use a variety of sizes of plastic bags for all our packaging. There are many options that you may prefer including corn-based eco-friendly composites and harder plastic clamshells that offer better protection from damage (great for living butter lettuce).

Wholesale products usually don't require packaging unless the buyer is retailing the product (wholesalers are usually middlemen). Head lettuce sold at market doesn't require bagging, although bags can be effective to protect the heads and to help prevent drying out on hot days. We prefer to display them live in water. Many customers will request a bag.

We have found that certain products need to be bagged just to prevent them from drying out even in mild weather. These include watercress, salad mix and mustards. Leave bags of cress and mustards open and folded down to create an attractive display (as long as they don't wilt).

Bags of salad mix must be sealed to ensure a long shelf life. Once a lettuce is cut, the exposed cut will undergo enzymatic oxidation which turns the base of the leaf an unattractive brown. This is the same process that occurs when an apple is cut. It is harmless but unattractive and therefore undesirable. Enzymatic browning can be delayed by chilling the product to slow the reaction and by sealing the product in a bag to reduce available oxygen.

Bags can be sealed with twist ties or elastic bands, but we highly recommend using a tape bander which is faster and more economical.

A plastic bag bander seals and trims the bag.

Harvest Cleanup

Harvest cleanup is a simple job and takes place after harvest. Move all freshly harvested (empty) rafts outside for cleaning and place blank rafts (with no holes) into the troughs. It is important to not leave troughs exposed to sunlight for more than a few hours as this will destroy the biofilm on the liner and encourage the growth of algae. We recommend having blank rafts (without holes) on hand for this purpose. If you don't have blanks, put cleaned rafts back in with empty net pots to reduce light penetration.

Use a hose with a strong nozzle to rinse off both sides of the rafts. A quick rinse will usually suffice to remove the debris — mainly dead leaves and crop residue — that has collected on the top. If needed, gently scrub the top with a soft bristle brush. If there was mold or mildew present in the harvested crop, sterilize the top of the raft with a weak vinegar or hydrogen peroxide solution of 3–5 ml/L. If using vinegar, rinse rafts well to prevent it from entering the system water. Stack rafts for replanting. Be careful to not bring in outside debris on the rafts.

IMPORTANT

Never scrub the biofilm on the underside of the rafts. A spray from the hose should suffice to remove any debris. This film is part of the biofilter.

Raft cleaning tools: hose with nozzle and soft bristle brush.

Net pots need to be washed between crops to remove roots and debris and to sterilize them if there was mold present. We recommend installing a washing machine outside the greenhouse for this purpose. Ours is located on the north side of the cooler, west of the Germination Chamber. Wash the pots on a gentle cycle using a small amount of non-toxic, biodegradable, unscented soap, or hydrogen peroxide if sterilizing for mold (15 mL of 30% per liter). After cleaning, nest the pots and store them in a clean bin until they are needed for transplanting.

The Production Cycle

So far in this chapter we have described the life of a plant from seed to harvest. In operational reality (from the perspective of the farmer), the entire production cycle happens in reverse order.

Your objective is to be operating at full capacity at all times: full Germination Chamber, Seedling Table and troughs. Once the system is at capacity, the first step in the production cycle is harvest. Harvest creates room in the troughs for seedlings to be transplanted, which creates room on the Seedling Table for new sprouts, which creates room in the Germination Chamber for new seeds.

The production cycle must be conducted in this order:

1. Harvest (including cleanup).
2. Thinning and spacing.
3. Rearranging rafts (conveyor cropping).
4. Transplant from Seedling Table to rafts and direct seeding.
5. Move trays from Germination Chamber to Seedling Table.
6. Plant new seeds into Germination Chamber.

Nutrient Deficiencies

Identifying deficient nutrients is a bit of an art. With some experience, it will be second nature. As with pest and disease management, your daily and weekly inspections are the primary diagnostic tool. Water testing for nutrient content will often further clarify the deficiency.

In the color photo section of this book, you will find a Nutrient Deficiency Key (created by Bright Agrotech; used with permission). There are numerous similar keys, but this is the best one we have seen. Follow the step-by-step flow to determine the deficiency. Additional copies of the key are available from Bright Agrotech via their website.

To resolve specific deficiencies, take the following actions. With any corrective action, run the appropriate chemical test(s) regularly afterwards to gauge the result.

Nitrogen

You should never have a lack of nitrates. Aquaponic systems produce large quantities of nitrates as the primary mineral.

Nitrogen deficiency shows as chlorosis (yellowing) across entire leaves. Note that some yellowing on the oldest, biggest leaves can be normal and should not immediately be assumed to be a nitrogen deficiently.

Nitrogen deficiency is caused by low feed input. Gradually increase the quantity of feed, test regularly for ammonia and nitrites to confirm the biofilter is not overwhelmed.

Calcium and Potassium

Calcium and potassium (along with iron) are the most commonly deficient minerals in our experience. Calcium and potassium must be relatively balanced. An excess of either can cause the other to be locked out (not available for uptake).

IMPORTANT

The deficiency symptoms for both calcium and potassium can look very similar to each other and can look similar to certain types of pest damage. If you feel one is deficient, run chemical tests to confirm there is a deficiency. Don't take action before testing as you may add to the problem.

Calcium levels will vary greatly depending on your source of water (we have very soft water with very low calcium). Sustained high humidity levels (over 80%) can also prevent calcium uptake due to decreased plant transpiration. To increase the calcium level, agitate the CFB sock more vigorously and/or more frequently. Increase sock use very gradually.

You can also supplement calcium by using an organic Cal-Mag solution as a foliar spray. There are numerous products available. We use Botanicare Cal-Mag+.

The most common cause of potassium deficiency is a lack of balance in pH management. You are likely shaking the sock too hard and/or often, which means the pH dosing controller is not inputting potassium hydroxide at sufficient levels. To remedy, cease sock shaking. Test daily and recommence sock shaking less frequently and/or aggressively once the two levels are balanced.

Iron

Iron will be lacking in almost all aquaponic systems. It is not contained in large amounts in the feed and will not likely be present in your source water in a usable form. Iron is highly reactive in aerobic environments and tends to combine with other molecules, becoming unavailable for absorption by the plants. This is less of an issue at a lower pH, a benefit of operating the system at the recommended pH of 6.5. Iron is a vital mineral for plant growth and your plants will almost certainly require more than is found in the water without supplementation.

In an aquaponic system, iron is usually present in two forms: ferrous iron (Fe^{2+}) which is soluble in water and available to the plants, and ferric iron (Fe^{3+}) which is insoluble in water and unavailable to the plants. Rust, or iron oxide, is ferric (insoluble) and should never be added.

We recommend supplementing iron monthly with 2–3 ppm of a high-quality chelated iron, whether your plants show signs of deficiency or not. Chelated iron is ferric (insoluble) iron that has been bound to an organic molecule to make it soluble and thus available to plants. The three most common forms of chelated iron are Fe EDTA, Fe EDDHA and Fe DTPA.

Use only Fe DTPA as the other two are either slightly toxic or ineffective below a pH of 7.0. We recommend Amazon Iron from Tailored Aquatics.

If an iron deficiency occurs, gradually increase the frequency of input.

High Ammonia

A final note that is not included in the Nutrient Key: a sign of a high ammonia level can be a white residue that looks similar to mineral scale around the outer edges of leaves. If you spot this symptom, do a chemical test for ammonia immediately. If the test confirms elevated ammonia, reduce or eliminate feed until the biofilter catches up.

Pythium damaged plant (middle). The first symptom is often a wilted plant surrounded by healthy plants.

Plant Diseases and Pests

THERE ARE NUMEROUS DISEASES AND PESTS that may impact your plants. We will discuss those we have found to be most common in our DWC system.

Pythium

Pythium is the most common and persistent pathogen we encounter. It is an opportunistic fungi that takes advantage of plants weakened by heat, low oxygen or pest damage, causing a disease known as "stem rot," "root rot" or "damping off."

Pythium spores exist nearly everywhere in soil and water but do not typically attack plants unless plants are stressed or there is physical damage. Fungus gnats are known carriers of pythium spores, and their larvae can spread infection by feeding on roots. Once inside the plant, pythium feeds on tissue and eventually releases spores back into the system to spread and infect other plants.

Typically, the first observable symptom of a pythium infection is poor growth or wilting for no apparent reason. Upon closer inspection, the plant will usually have a rotten brown stem and in advanced cases will separate from the root ball when lifted.

Pythium prefers dark, wet anaerobic zones. Placing one air stone under each raft per our design will greatly reduce the potential for anaerobic zones near plant

Pythium damage causes stem rot. The head of this lettuce is fully separated from the root ball.

Powdery mildew.

roots. Pythium spores are also destroyed by UV doses in the range of 30–40 mJ/cm^2, well within the sterilizing capabilities of the recommended UV system.

Inspection and early removal is the primary management method, along with planting only healthy, vigorous seedlings. Oxygen-saturated water and lower air humidity are also beneficial, as is watering plants with system water as described in Chapter 9.

If pythium is rampant and spraying is required, use a bio-fungicide product such as *Natria* which is non-toxic to fish and OMRI listed. Sprays should only be used as a last resort in cases of severe infection. See Pesticides and Fungicides later in this chapter.

Powdery Mildew

Powdery Mildew (PM) is a fungal disease that affects a wide range of plants and is common in greenhouses. It is easily identified: it looks like a thick coating of icing sugar. It often starts as small circular outbreaks before spreading across a leaf or stem. PM can be a problem at almost any time of year, depending on local conditions.

The primary management method is environmental control to prevent conditions which encourage PM growth, and inspection and early removal of infected plants. Healthy plants, inspection and early removal are the first and best remedies. Carefully remove plants from the greenhouse, taking care not to spread spores and move to compost.

The powder that is visible on a plant is only the fruiting section of the fungus which penetrates the leaves. Once a plant is infected, the PM cannot be fully removed without toxic systemic pesticides, but the recommended treatment will prevent it from spreading to other plants.

PM can be treated by spraying high pH water (9.0–10.0) with 3 mL of 30% hydrogen peroxide per liter. Note that higher concentrations of peroxide will burn or whiten plant leaves. Both high pH water and H_2O_2 combat PM and in small quantities can be used without concern for your fish. Applications of high pH water with H_2O_2 may be required every 2–3 days as new fruiting bodies emerge.

Some fungicides, such as Natria, specify that they are effective against PM, but as with the H_2O_2 spray, it only destroys the fruiting body and not the mycelium which penetrates the leaf tissue. High pH water with H_2O_2 is the best option if spraying is required.

Some seed producers are now offering seeds that are resistant to PM via selective breeding. We recommend such seeds if they are available.

Fungus Gnats

Fungus gnats are a common pest in almost all greenhouses. In their adult form, they look like very small mosquitoes that hover in swarms 1–2′ above the plant canopy. They lay their eggs in the soil around a plant's base. The resulting larvae look like small (1–5 mm) white worms with black heads. The larvae feed on plant roots until they pupate and fly above the canopy.

The destruction of roots is typically not a big problem as long as plants are healthy and vigorous and the gnat population is not extreme. The bigger problem is that gnats can also carry and transfer pythium, which is far more dangerous.

Gnats are very difficult to eradicate. Limiting their presence should be your goal. The most effective control is planting only strong, vigorous seedlings. Strong plants will outgrow direct gnat damage and will greatly minimize the ills of pythium.

There are numerous biological controls that are effective in killing gnats in the larval stage via application to the soil. However, due to the small soil content and constant soil wetness in a DWC system, such controls are not very effective and likely not worth considering. Similarly, using a pesticide spray or fog to kill adults on contact in the air is only temporarily effective as the larvae remain unharmed. Sprays should be used with caution and only in extreme cases.

Aphids

Most common in hotter seasons, aphids are another ubiquitous greenhouse pest. Typically lime green or brown, aphids reside on plant leaves and shed a white husk. Both pest and husk are easy to identify. They spread very quickly and don't do a lot of damage unless present in very high numbers.

Aphids.

The biggest problem with aphids is not necessarily the damage they do but their impact on sales: customers do not like to see aphids on plants, and restaurants don't like having to wash produce multiple times to remove them.

Options for aphid management include washing them off and/or drowning them by spraying plants with system water, sprinkling diatomaceous earth on the plants to dessicate the aphids, and using a pesticide spray. As always, pesticides should be used with caution only in extreme cases.

If you find groups of aphids on your plants at harvest, they can also simply be washed off. It may take two or more rinses. Aphids die off naturally in colder seasons.

Cabbage Loopers

Loopers are small green inch-worm pests that are typically less than 1 cm long and have a voracious appetite. They are mostly nocturnal feeders and can ruin multiple plants overnight. The first sign of loopers is typically a patchwork of holes in leaves accompanied by large quantities of small green balls of feces.

Loopers are best managed via the same controls as aphids: direct spraying with system water, sprinkling diatomaceous earth or, as a last resort, a fish-safe pesticide.

Earwigs

Earwigs prefer decaying organic material but will also feed on tender lettuce leaves, chewing small holes similar to loopers. Earwigs are nocturnal feeders that do less leaf damage than loopers. They nest in the base of a plant, making them hard to see on casual inspection. Visual indicators of an infestation are small holes chewed in leaves accompanied by necrotic patches on tips of leaves on which the bugs have laid eggs at the base.

Our experience is that earwigs prefer a plant that layers its leaves and forms a protective shelter for the bug. Romaine and butter lettuce seem to be their favorites.

Left unchecked, earwigs will quickly spread and damage plants to the point where they are no longer marketable. Inspection and early removal is key to managing them. Use diatomaceous earth to control outbreaks and, as a last resort, a fish-safe pesticide.

Earwig.

Pill Bugs

Pill bugs are actually crustaceans (like lobsters) and breathe with gills instead of lungs. They prefer very damp and dark locations and are nocturnal feeders. Likely found in and around the net pots, they typically eat decaying leaves and root tissues but will occasionally eat tender seedlings or young leaves that are touching the ground. Damage from pill bugs is usually minimal and they are typically only detrimental to market appeal.

As always, inspection and early removal is key to keeping them under control. Outbreaks can be controlled with applications of diatomaceous earth and, as a last resort, a fish-safe pesticide.

Slugs

Slugs are a fact of life for any humid, moist or wet plant growing space, including all greenhouses. They are best managed in small numbers by manually removing and drowning them.

Beer traps, coffee grounds and copper are all purported to be natural solutions, but we have found none of these to be even moderately effective. We have watched them drink the beer and move on, slide over coffee grounds, and show no sign of repulsion when they encounter copper wire or tape.

Crushed oyster shells or diatomaceous earth, commonly available in large bags for minimal cost, are moderately effective as a physical barrier if laid in an unbroken ring around the troughs. This doesn't kill slugs but provides a barrier due to sharp edges. As a last resort, iron phosphate bait pellets can be used with caution.

Birds

Greenhouses can be ideal refuges, playgrounds and nesting areas for birds. In summer when the walls and doors remain open, the greenhouse can act as a huge flying insect trap, and birds love to take advantage of the easy buffet. In winter when food and warmth are scarce, birds will take up residence and eat any newly germinated seeds that are left uncovered. Unfortunately, despite their benefits as insect eaters, birds should be discouraged from entering the greenhouse as they tend to leave droppings on the crops. As much as we love birds, we chase them out using a fishing net before they get used to inhabiting the space.

In summer when the roll-up sides are left open, birds can be excluded by securing shade cloth or other mesh screen across the opening.

Rats

Rats can destroy whole trays of young seedlings rapidly, like a drive-through salad bar. Additionally, rat feces is to be avoided due to health risks. Rats are primarily attracted to the smell of the fish feed and will make themselves at home if they can access it. Keep your fish feed safely locked in vermin-proof storage and take care not to spill pellets on the ground when feeding.

Manage rats with traps and/or a farm dog. Be mindful of where you lay traps to avoid injury to workers, children and pets. Do not use rat poison. Poison has no place in or near an aquaponic system.

A dead mink and our farm dog, Tango, defender of trout.

Mink and Marten

Mink, marten and other semi-aquatic mammals in the weasel family may view your farm as a restaurant. Attracted by the smell, they will dive into a tank and kill numerous fish before taking just one. Mink are particularly fearless and will not be deterred by human activity alone.

The easiest control is the presence of a farm dog for whom hunting mink and other small mammals will be a desirable job. Live traps are effective when baited with fresh fish, though you must then relocate the animal. If a lethal trap is used, be mindful of children and pets. Do not use poison.

Bears

Bears are not common in our experience despite living in an area where they roam. As opposed to mink, bears are usually very wary of humans. The lights, sound and human activity of an aquaponic farm will usually keep all but the most desperate bears away. It is beyond our capacity to provide safe instruction for an encounter with a bear. If you have many bears in your area or are worried about a bear encounter, research your plan of action with a bear expert.

Plant Disease and Pest Management

Plant disease and pest management options are very restricted in an aquaponic system due to its biological complexity and interconnectivity. A varying percentage of any chemical or substance used will end up in the system water and can easily be toxic to the fish.

IMPORTANT

Pesticides and fungicides, even those that are derived from natural or botanical sources and those that are OMRI listed, may be toxic to your fish even if considered safe for humans or other animals. Extreme caution must be used with any substance sprayed or otherwise applied on or around your plants.

We highly recommend learning and implementing Integrated Pest Management (IPM) protocols. The United Nations Food and Agriculture Organization defines IPM as "the careful consideration of all available pest control techniques and subsequent integration of appropriate measures that discourage the development of pest populations and keep pesticides and other interventions to levels that are economically justified and reduce or minimize risks to human health and the environment. IPM emphasizes the growth of a healthy crop with the least possible disruption to agro-ecosystems and encourages natural pest control mechanisms."

The first and best lines of defense against pests and disease are cultural and environmental controls. We cannot overstate their importance in preventing and minimizing disease and pests.

Cultural Controls

Cultural controls are practices that reduce pest establishment, reproduction, dispersal and survival. The most basic and important cultural control is the daily and weekly plant inspection. Diseased or pest-infested plants should removed from the facility immediately. Sterilize your hands after removing plants to prevent the spread of disease.

Plant quality and maintenance is also a cultural control. If a particular plant type does not do well in your system, do not grow it. In an aquaponic farm, the plant must match the system, not vice versa.

Every week you will germinate more seeds than you will require, for one primary reason: so that you always have healthy, strong, vigorous seedlings to transplant into the troughs. Do not transplant weak seedlings. Thinning and/or spacing plants so they are not crowded, stressed and stretching for light helps to keep plants strong.

IMPORTANT

It is often better not to plant a full crop than to plant weak seedlings that can lead to disease or pest outbreaks that may also infect other crops. More than any other single factor, having strong plants and a healthy ecosystem is your best defense against disease and pests.

In hotter/drier seasons, it is beneficial to spray plants via your diffused water wand with system water once or twice a week. This assists in washing off pests and acts as a foliar feed which can be particularly useful for calcium and potassium uptake.

Environmental Controls

Environmental controls are the conditions of your greenhouse: air and water temperature, humidity, light level and air movement. In order to have strong plants, you must have a good environment.

Water temperature should be automatically controlled by the heat pump and kept at a consistent 15–17°C (59–63°F) all year round. Air temperature should never drop below 5°C (41°F) including at night. Set the propane heater to turn on at 5°C; place the thermostat as near to the plant canopy as possible. Air temperature less than 7°C (45°F) will greatly reduce pest populations: fungus gnats, aphids and earwigs will die off or hibernate. Air temperature above 25°C (77°F) should be reduced via passive and/or powered ventilation and/or shade cloth.

In cold seasons, use circulation fans to equalize temperature and humidity throughout the greenhouse. In warm seasons, these fans should be turned off to allow the air to stratify. With the fans off, warm air will rise to the roof where it can be vented out, drawing cooler air in through the doors and roll-up sides.

For optimum plant growth, humidity should be 50–80% year-round. Hygrometers are inexpensive and commonly available. Place them near the plant canopy in various locations as well as outside. PM and most other fungal diseases thrive in high humidity. Severe outbreaks can be prevented by lowering the humidity: increase air movement via circulation fans and/or increase ventilation via roll-up sides or peak vents. Large dehumidifiers designed for greenhouses are effective but draw huge amounts of electricity.

Heating/Ventilation Cycle

In winter, if the outside humidity is lower than inside the greenhouse (which is nearly all the time), humidity can be lowered manually with an active heating/ventilation cycle.

1. Use the propane heater to temporarily increase the temperature by 5 degrees or so. (Warm air holds more moisture than cool air.)
2. Wait at least 15 minutes for the air to circulate and absorb moisture via circulation fans, then turn off the heater and fans.
3. Turn on the power ventilation system to vent the warm moist air. Monitor temperature and humidity throughout the cycle.
4. Once humidity has dropped, turn the ventilation system off and circulation fans back on. Reset the propane heater to engage at 5°C.

This cycle is most effective at dawn and dusk when humidity in the greenhouse tends to be the highest. A few cycles may be needed to lower humidity to acceptable levels. The only drawback is using additional propane.

Sticky Traps

Sticky traps are cheap, bright yellow paper cards covered in a sticky resin that are very useful in helping to identify early stages of pest infestation. We recommend hanging them just above the canopy at random locations to identify when and where a pest population is growing, by observing the concentration of trapped insects on each card.

Yellow sticky traps.

Predatory Insects

Predatory insects target pest species. A common example is using ladybugs to control aphid populations. We have experimented with numerous predatory insects and found them to be helpful but not capable of resolving major pest issues. They can also be quite expensive. Predatory insects are best used in closed greenhouses (when the sides are closed) to combat low-level pest infestations. We have found them to be ineffective in major outbreaks.

Using Pesticide and Fungicide Sprays

The term "pesticide" is much maligned and often misunderstood. A pesticide is any substance used for destroying insects or other organisms harmful to cultivated plants or animals. These include synthetic chemicals as well as naturally derived compounds and minerals, botanically sourced substances and even biological pesticides which contain organisms that are pathogenic to the target pest. Many pesticides and fungicides are OMRI listed or approved for use in organic production because they are considered safe for humans and have little negative effect on a natural environment when used as specified.

Pesticide use in an aquaponic greenhouse is inevitable, as cultural and environmental controls won't resolve severe pest outbreaks and commercial growers do not have the luxury of experimenting with marginally effective homemade remedies. Only pesticides that are OMRI listed or allowed under organic certification standards should be considered.

One commonly used botanical pesticide approved for organic production is pyrethrins. Derived from the chrysanthemum plant, it is highly effective against many common pests, including fungus gnats, aphids, loopers and earwigs. It is also highly toxic to fish. Substances such as pyrethrins should only be used in the most extreme outbreaks when all other methods have failed and only if the level of toxicity to the fish is calculated and a maximum application limit strictly adhered to.

IMPORTANT

A pesticide that is considered safe for human consumption is not necessarily safe for fish.

Lethal Concentration Calculation

Before using any substance as a plant spray, you must first determine whether it is toxic to your fish and, if so, whether it can be applied in a manner that is safe for the fish, yet still effective. All pesticides and many other chemicals and agricultural products have been tested on a variety of organisms to determine their potential toxicity, which is quantified as the lethal concentration (LC_{50}) or lethal dose (LD_{50})

Generally, the LC_{50} is the concentration (in water) of a substance that will kill half of the tested species over a period of time (usually 96 hours). However, it is not an exact science because there are many factors that can aggravate or mitigate the toxicity of a substance, including the amount of diluted spray that actually enters the water, the length of exposure time, the substance's rate of decomposition, water temperature and pH.

Determine the LC_{50} of the substance for trout (or your chosen fish species) using the MSDS of the product, the manufacturer's website or online resources such as the Pesticide Properties Database.

IMPORTANT

It is our opinion that a pesticide should only be considered for use in your aquaponic system if you can be certain that less than 10% of the LC_{50} of a substance will make its way into the water.

To calculate the LC_{50} for your system, you will need to know the amount of active ingredient in the pesticide, the LC_{50} of the active ingredient for trout (or your chosen species), the volume of water in the system and the amount of diluted spray that is needed to provide effective coverage.

The following is an example of a LC_{50} calculation for the pyrethrin-based product Pyganic™. This is for demonstration purposes only. We recommend you seek the advice of a pesticide professional prior to using any pesticide or fungicide.

1. Multiply the volume of water in the system by the LC_{50} of the pesticide to find the LC_{50} value for the system. The Pesticide Properties Database states that the LC_{50} of pyrethrins for rainbow trout is .005 mg/L, therefore:

$$\text{(system volume)} \times (LC_{50}) = LC_{50} \text{ of the pesticide}$$

$$\text{(66,000 liters)} \times (.005 \text{ mg/L}) = 330 \text{ mg or } 0.33 \text{ mL}$$

For a system that contains 66,000L of water, the LC_{50} of pyrethrin is 0.33 mL. This shows just how toxic pyrethrins are to fish.

2. Pyganic™ contains 1.4% pyrethrins. The LC_{50} for Pyganic™ in your system is therefore:

(LC$_{50}$ of the pyrethrins in your system) / (concentration of the product) =
LC$_{50}$ of product

(0.33 mL) / (0.014) = 23.5 mL of Pyganic™

Pyganic™ recommends a dilution of 1–2 ounces per gallon of water (7.5–15 mL/L) which would seem to allow only 3L of total diluted spray (at 7.5 mL/L), yielding a total pyrethrins load of 0.315 mL, to reach LC_{50}.

It is our opinion that a pesticide should only be considered for use if you can be certain that less than 10% of the LC_{50} of the pesticide will make its way into the water. In this example, that would mean less than 160 mL of mixed spray at the highest concentration or 320 mL at the lowest concentration.

We know from experience that, used cautiously and properly, up to 8L of Pyganic™ can be safely used in our design at 7.5 mL/L, as often as every four days, with no harm to fish, due to several mitigating factors, and that 8L of spray is approximately sufficient to cover the greenhouse.

3. Determine the amount of active ingredient that would be in the volume of diluted spray required for effective coverage. Assuming you require 8L of diluted pesticide to spray the entire greenhouse with Pyganic™ at the recommended concentration of 7.5–15 mL per L:

(15 mL per L) × (8L of spray) = 20 mL of Pyganic™, which contains:

(120 mL) × (1.4% concentration) = 1.68 mL of pyrethrins

From this we can see that 8L of mixed spray at the highest application rate contains five times the LC_{50} of pyrethrins. Even using the lowest application rate of 7.5 mL/L, it would still contain 2.5 times the LC_{50}.

Pesticide infiltration can potentially be problematic in DWC systems, as runoff from the plants can easily make its way between the rafts and into the water. For all pesticides and fungicides, it is imperative to avoid spraying the plants to the point of runoff. Applying the spray with a fogger and using a spreader/sticker additive will also greatly reduce runoff.

There are also other influencing factors that may mitigate toxicity. Remember that the median lethal concentration for trout is .005 ml/L *for a duration of 96 hours (4 days)*. When exposed to water and sunlight, pyrethrins degrade rapidly, having a half-life of 12 hours. This means that 50% of the pyrethrins will have decomposed 12 hours after application, and 50% of the remainder in another 12 hours, and so on. The UV sterilizers may also be of some benefit in this case, as pyrethrins are rapidly photolysed by UV light in the presence of water.

IMPORTANT

We do not specifically recommend the use of pyrethrins or any pesticide or fungicide. This section demonstrates how an OMRI listed pesticide that is considered "safe for humans" can be toxic to your fish. Always consult a pesticide professional prior to using any pesticide or fungicide.

If you do choose to use any pesticide or fungicide, be extremely cautious and follow these steps to prevent the spray from entering the water:

1. Use a fogger rather than a sprayer.
2. Add a spreader additive which will decrease the surface tension of the spray and prevent the formation of large droplets on the leaf surfaces.
3. Do not spray plants to the point of runoff. This is vital.
4. For three days after spraying, do not water plants with system water, as this will wash pesticides into the system. This is also vital.

Always read and follow all information on the product label. In many places, it is against the law to use a pesticide in any way other than as directed. Only spray during the morning or evening, never in full sunlight. Many chemicals can be phytotoxic to plants if applied in heat or intense sunlight.

IMPORTANT

Always read the MSDS of the specific product you plan to use. Remember that the toxicity to fish is likely not considered. Pay attention to "inactive ingredients" which may also be toxic to fish. Ensure the pesticide is legal to use on commercial food crops.

Table showing pesticides we have found to be effective on our farm.

Active Ingredient	Target Pests	Properties	Toxicity to Trout	Comments
Pyrethrins (botanical extract: chrysanthemum plant)	All soft-bodied insects, earwigs, pill bugs	Indiscriminate contact killer, nerve toxin	Highly toxic 96hr LC_{50} = .005 mg/L	Will also kill beneficial insects. Photolysed rapidly in water by UV.
Azadirachtin (botanical extract: neem tree seeds)	All soft-bodied insects, earwigs	Repellant, antifeedant, insect growth regulator	Moderately toxic 96hr LC_{50} = 2.25 mg/L	Will affect beneficial insects. Biodegrades rapidly in water.
BTi (biological: *Bacillus thuringiensis sp. Israelensis*)	Larvae of flies and mosquitoes, notably fungus gnats	Bacteria, releases protein that affects larvae of target species only	Non-toxic	Effective for gnat control if applied regularly to root balls.
BTk (biological: *Bacillus thuringiensis sp. kurstaki*)	Larvae of lepidopterans Butterflies and moths (including loopers)	Bacteria, releases protein that affects larvae of target species only	Non-toxic	
Bacillus subtilis (biological)	Fungal diseases	Bacteria, broad-spectrum biofungicide	Non-toxic	
Insecticidal soap (potassium salts of fatty acids)	All soft-bodied insects, earwigs, pill bugs	Destroys waxy outer layer on contact, desiccating target pest	Slightly toxic 96hr LC_{50} = 18 mg/L	
Diatomaceous earth (mineral: silicon dioxide)	All soft- and hard-bodied crawling insects	Microcrystalline silica powder, highly abrasive to insects, causes desiccation	Non-toxic. High concentrations may scar fish gills	Wear a respirator, harmful to lungs if inhaled

Re-entry Interval

The re-entry interval is the minimum length of time that must pass between the time a pesticide was applied to an area or crop and the time that people can go into that area without protective clothing and equipment. Every pesticide has a re-entry interval listed on the label that must be followed.

Personal Protective Equipment (PPE)

PPE must be worn anytime you are spraying, even if the product is considered safe for humans. Many "safe" pesticides can still cause irritation to skin or lungs. A respirator is crucial for protecting yourself.

PPE for spraying pesticides in the greenhouse include:

- Tall boots or shoes with socks
- Long pants that cover the ankles
- Long-sleeve shirt buttoned up to the neck
- Nitrile or other chemical resistant gloves
- A hat or hood and face shield or goggles
- A respirator

A set of chemical-resistant overalls are recommended as they offer more protection than clothing and can be worn over your work clothes.

Foggers

Foggers put out an atomized mist rather than a small spray of droplets. The fine mist created by a fogger is far less likely to form droplets which can fall past or off leaves. This is vital to minimizing runoff. We recommend using an electric fogger rather than a pressurized sprayer. There are numerous types and quality of foggers. We recommend Hudson fog atomizers.

Standard Operating Procedures and Protocols

THIS CHAPTER CONTAINS OUR TRIED AND TESTED Standard Operating Procedures (SOPs), logs and protocols to run the RCA system. Most of the details for each task have been covered earlier in the book. The purpose of this chapter is to detail when and at what frequency each task is performed.

Recreate each log and protocol and keep in the greenhouse, with logs to be checked off or filled in as they are completed. Keep a file for completed logs. This record of activity is standard good practice and will allow you to examine the history of your farm.

Sample logs and protocols are at the end of the chapter. Formatted blank logs are available on our website.

This chapter contains:

- Daily Log
- Weekly Tasks (Phases 1–4)
- Weekly Seeding Chart
- Crop Log
- Fish Sample Log
- Cohort Log
- Monthly and Seasonal Tasks
- Power Outage Protocol (with Cistern)
- Water Flow Alarm Protocol

Daily Log

These tasks should take very little time but must be done every day or more than once per day.

Weekly Tasks

Weekly tasks constitute the majority of work on an aquaponic farm. These include plant production, fish sampling, water testing and cleaning. Tasks are broken down into four phases, each performed once per week in a single day or spread over additional days if required. You can choose the day of the week each phase will take place, but we recommend setting a regular schedule.

If you have a primary market day (your biggest selling day, often a farmers market), Phase 1 is completed the day before. Market days are usually long, thus little work is done in the greenhouse other than daily tasks. Assuming one day for Phase 1 and one day for market, Phases 2–4 are done over the next five days. Specific tasks may vary if you grow different plants. Expect Phases 1–3 to be your longest work days.

Notes for Phase 3

To clean the heat pump filter:

1. Close the inlet and outlet valves and shut off the unit's water circulation pump.
2. Disassemble the filter housing and remove the filter. Clean the filter with a hose.
3. Reassemble the unit. Plug in the water pump and open the inlet and outlet valves.

To clean the water pump intake:

1. Shut off the pump, then close the valve (above the pump).
2. Unscrew the suction pipe (below the pump).
3. Inspect the pipe and pump impeller for debris. Backflushing may be useful to dislodge debris.
4. Reassemble the suction pipe and open the valve before turning on the pump.
5. Confirm the pump is functioning.

Scrub either Tank 1, 2 or 3 each week. Rotate so each tank is cleaned every three weeks. Additionally, scrub the drain screens of all tanks every week, then use the cleaning brush to remove the drain screen and push any debris into the MWP. Reseat the drain screen. Pop the SPA on each tank to flush debris down the MWP. Use only the dedicated brush for each tank.

To clean the CFB:

1. Remove the seven outermost screens from the upstream side and the one outermost screen from the downstream side, leaving only the two innermost screens that contain the MBBR media.

2. Use a hose with good pressure to rinse each screen into the Waste Tanks. Gently tapping the screens will help to dislodge accumulated solids.
3. Replace the screens back into their respective slots.
4. Remove the two innermost screens containing the MBBR media and clean.
5. Replace the innermost screens, moving all MBBR media from outside the containing screens back into the central MBBR area.
6. Replace the lid.

To clean trays, hose them off then agitate in a bin of soapy water (non-toxic, biodegradable, unscented soap). Rinse trays well, then lay out to dry. Once dry, stack in an offset pattern to avoid having them stick together. Store stacked trays under the workbench.

Weekly Seeding Chart

We find it is useful to have an erasable chart to note what is to be seeded. This is particularly helpful if the person doing the seeding is not the person deciding what is to be seeded. To make the chart erasable, laminate it and use dry erase pens.

Crop Log

A Crop Log keeps track of the movement of a single plant variety over several crops as they mature. Note only one plant variety per Crop Log. Include a marker with plant variety and seed plant date on each tray and on each raft.

Fish Sample Log

The Fish Sample Log is used to record sampling data and calculate the average weight per fish. Sample one tank every week so that each tank is sampled every three weeks. Copy the relevant info from the Fish Sample Log to the appropriate Cohort Log.

Cohort Log

The Cohort Log is used to track the number of fish, total biomass, daily feed rate, mortality and disease incidences, treatments and transfer dates for each cohort. Each tank containing a cohort of fish must have an easily accessible Cohort Log.

Record every type of event that occurs, such as a mortality, treatment, cohort sampling or tank transfer. Update the number of fish in the cohort and recalculate the tank biomass and daily feed rate every time a fish is removed for any reason.

Monthly and Seasonal Tasks

These tasks are less frequent but must be done regularly to ensure consistent high productivity. We recommend putting repeating dates on your calendar.

Replacing UV Bulbs

UV bulbs will typically need to be replaced every year (after 9,000 hours of operation). After one year, light output will diminish below 90% of the bulb's rating, decreasing its germicidal effectiveness.

Each UV lamp is contained within a clear quartz sleeve to protect the bulb from direct contact with water, allowing it to operate at high temperature without warming the passing water. Changing the bulb is simple, as it is not necessary to shut off the water flow or drain the unit.

When installing a bulb, do not contaminate the bulb surface with dirt or oils from your hand as contamination can reduce its lifespan. Inspect bulb prior to installation to confirm it is faultless, both in manufacture or due to contamination. Carefully wipe the length of each bulb as you install it.

IMPORTANT

If for any reason the main pump needs to be temporarily shut down, also shut down the power to the UV units. Leaving the bulbs powered with no water flow will rapidly increase the water temperature in the units which can degrade the life of the bulb.

DAILY LOG

RAINCOAST
AQUAPONICS

Date: _____ Nov 15, 2017 _____

Water Temp: _____ 16° _____

Indoor Temp: ___ 18° ___ (high) / ___ 7° ___ (low)

Outdoor Temp: ___ 10° ___ (high) / ___ 3° ___ (low)

		MORNING	MIDDAY	EVENING
Record pH (normal is 6.5–6.6)		pH: **6.5**		pH: **6.5**
Plant observation		✓		
Fish observation		✓		
Feed Fish: see Cohort Logs for daily feed rate - spread daily feed rate across 2 or 3 feedings - one scoop is **100** grams	Tank 1 (grams)	250	250	250
	Tank 2 (grams)	600	600	600
	Tank 3 (grams)	1000	1000	1000
Shake calcium sock (every 1-3 days)		✓		
Clean trough standpipe filters		✓		✓
Water seedlings		✓		✓
Flush (pop) SPAs for Tanks 1-3		✓	✓	✓
Check sump level and fill as needed		✓	✓	✓
Check CFB filters and clean as needed		✓		
Check pest traps		✓		

Notes:

PHASE 1
HARVEST & PACKAGING

Date: _____

HARVEST .... ☐
completed

- Harvest head lettuce into bins (optional: with the roots on in ½" water).
- Harvest romaine into full heads and hearts.
- Harvest living butter lettuce: remove hanging roots and net pot and wrap root ball in bag with a rubber band.
- Harvest watercress: leave or remove lower roots, depending on if sold separately.
- Harvest mustards by removing individual leaves.
- Harvest baby kale by mowing, leaving 2"-4" to regrow.
- Rinse plants as needed.
- Label bins and move to cooler.
- Record harvest totals on Crop Log.

SALAD MIX .... ☐
completed

- Rinse each bin of produce prior to assembly.
- Cut bases off lettuce, move to clean bin.
- Layer mix in clean bins.
- Rinse completed mix. Spin dry.
- Place dried mix in clean bins.
- Label bins and move to cooler.

PORTIONING AND PACKAGING .... ☐
completed

- Handle all plants gently when bagging.
- Move labeled bins to cooler.

HARVEST CLEAN UP .... ☐
completed

- Move rafts outside and clean. Do not scrub biofilm.
- Insert blank rafts into troughs.
- Wash net pots in washing machine.

Notes:

PHASE 2
ROTATION, INSPECTION & TRANSPLANT

Date: _____

THINNING AND SPACING .... ☐
completed
- Thin and space plants as needed.

QUALITY CHECK ALL TROUGHS .... ☐
completed
- Visually check every plant in toughs for weak and diseased.
- Remove any culls from greenhouse immediately.
- Place used net pots in dirty bin to be washed.
- Condense crops as needed. Replace with blank rafts.

CROP ROTATION .... ☐
completed
Move rafts so older plants are downstream, leaving space upstream for transplants.

TRANSPLANT SEEDLINGS .... ☐
completed
Use the most mature seedlings.
- Avoid weak seedlings or those with signs of disease.
- Place transplanting media into net pots.
- Use transplant tool to remove seedlings from trays and seat in net pots.
- Place net pots into a clean raft.
- Place raft into upstream end of troughs.

Notes:

PHASE 3
SEEDS, SEEDLINGS & CLEANING

RAINCOAST
AQUAPONICS

Date: _____

GERMINATION TO SEEDLING TABLE .... ☐
completed

- Move trays from germination chamber to seedling table.
- Place new trays directly under lights.
- Leave domes on new trays for 2 or 3 additional days.
- Remove weak or diseased seedlings.
- Thin to one seed per cell, as needed.
- Record on Crop Log.

SEEDING TRAYS .... ☐
completed

- Fill workbench with trays.
- Top up all cells with seedling media. Water trays.
- Fill all cells with seedling media again (do not water).
- Plant seeds and move to germination chamber.
- Record on Crop Log.

DIRECT SEEDING .... ☐
completed

- Refer to Weekly Seeding List.
- Place transplanting media into net pots.
- Place germination media over transplanting media.
- Place net pots into a clean raft.
- Sprinkle seeds into net pots.
- Place raft into upstream end of trough.
- If needed, cover raft with clear plastic.
- Record on Crop Log.

CLEANING (in this order) .... ☐
completed

- Clean trough drain screens.
- Clean heat pump filter.
- Clean water pump intake.
- Scrub Tank 1, 2 or 3 (rotate weekly). Tank cleaned: _____
- Scrub drain screens on all tanks.
- Scrub under drain screens on all tanks.
- After scrubbing, pop SPAs on all tanks.
- Scrub tank manifold if needed.
- Pop the RFS, then scrub, then pop again.
- Clean the CFB.
- Clean all dirty totes with soap and water. Rinse well.
- Clean seedling trays.

Notes:

PHASE 4
FISH SAMPLING AND WATER TESTING

Date: _____

FISH SAMPLING .... ☐
completed

- Sample Tank 1, 2 or 3 (rotate weekly). Tank sampled: _____
- Record sample on Fish Sample Log. Transfer mean average weight to Cohort Log.
- Complete Cohort Log.

WATER TESTING .... ☐
completed

pH: _____
Ammonia: _____
Nitrites: _____
Dissolved oxygen: _____
Take samples from a tank inlet spout.

Nitrates: _____
Calcium: _____
Potassium: _____
Iron: _____
Take samples from a trough inlet spout.

Source water pH: _____
Source water hardness: _____
Take samples from source water (before it enters system).

Confirm chemical pH is the same as digital pH.

WASTE MANAGEMENT .... ☐
completed

- Draw solids from settling tank to fermentation tank.
- Drain supernatant from settling tank.
- Refill settling tank from Waste Tanks.
- If fermentation complete in fermentation tank, bottle for use.
- Recycle fertilizer back into system (optional). Amount added: _____

Notes:

WEEKLY SEEDING CHART

PLANT VARIETY	# OF TRAYS OR RAFTS
Tropicana	5 Trays
Coastal Star	5 Trays
Alkindus	5 Trays
Breen	3 Trays
Watercress	6 Rafts
Mizuna	2 Rafts
Abundance Kale	3 Rafts

Notes:

CROP LOG

RAINCOAST
AQUAPONICS

Plant   *Tropicana*

SEED PLANT DATE	# of TRAYS PLANTED	SEEDLING MOVE DATE	TRANSPLANT DATE	# of RAFTS PLANTED	HARVEST DATE	HARVEST YIELD	NOTES
Nov 6	4	Nov 13	Dec 4	11	Jan 6	362	
Nov 13	4	Nov 20	Dec 11	12	Jan 14	300	Pythium issues this crop
Nov 20	4	Nov 27	Dec 18	11	Jan 20	380	

The Aquaponic Farmer

FISH SAMPLE LOG

RAINCOAST
AQUAPONICS

Date: _Aug 12_   Cohort: _10_

SAMPLE #	TOTAL KG	# FISH	AVERAGE PER FISH (GRAMS)
1	3.51	6	585
2	2.94	5	588
3	2.93	5	586
4	3.50	6	583
		MEAN AVERAGE:	586

Date: _____   Cohort: _____

SAMPLE #	TOTAL KG	# FISH	MEAN AVERAGE PER FISH (GRAMS)
1			
2			
3			
4			
		MEAN AVERAGE:	

Date: _____   Cohort: _____

SAMPLE #	TOTAL KG	# FISH	MEAN AVERAGE PER FISH (GRAMS)
1			
2			
3			
4			
		MEAN AVERAGE:	

COHORT LOG

Cohort# _____ 10 _____

DATE	EVENT	TOTAL FISH	MEAN AVERAGE PER FISH (GRAMS)	TOTAL BIOMASS	FEED RATE %	DAILY FEED (GRAMS)
Sept 7th	Cohort Introduction	250	25g	6.25kg	2%	125
Sept 9	Mortality −3 Stress	247	25g	6.17kg	2%	123.5
Sept 30	Sampling	247	37.5g	9.25kg	2%	185
Oct 25	Sampling	247	61g	15kg	2%	300
Nov 14	Sampling	247	98g	24.2kg	2%	484
Nov 20	Mortality −4 Predation	243	98g	23.8kg	2%	476
Dec 8	Sampling	243	137g	33.3kg	1.5%	500
Dec 30	Sampling	243	166g	40.3kg	1.5%	605
Jan 10	Sampling/ Transfer Tank2	243	220g	53.5kg	1.5%	802
Jan 20	Mortality −4	239	220g	52.6kg	1.5%	790
Jan 21	Mortality −6	233	220g	51.3kg	1.5%	770
Jan 22	Mortality −5/ H2O2 treatment	228	220g	50.2kg	1.5%	753
Feb 10	Sampling	228	273g	62.2kg	1.5%	934
March 2	Sampling	228	345g	78.7kg	1.5%	1180
March 24	Sampling	228	400g	91.2kg	1.5%	1370
Apr 16	Sampling	228	512g	116.7kg	1%	1167

MONTHLY AND SEASONAL TASKS

Monthly Tasks

- Calibrate pH controller. Check with chemical pH test.
- Test monitoring and backup oxygen systems: close water pump valve to confirm monitoring.
  system engages and notifies you and that backup oxygen system engages.
- Change light timers over troughs and seedling table as needed.

Seasonal Tasks

- Replace UV bulbs (once per year).
- Service propane heater (once per year in the fall, before it is used that season).
- Clean heat pump exchanger (minimum twice per year).
- Install or remove shade cloth.
- Turn circulation fans off in summer, on in winter.
- In spring and fall, raise and lower roll-up sides daily; in summer, leave raised.
- Replace greenhouse poly covering (every 4-5 years).
- Replace HID bulbs over troughs and seedling table (every 5-10 years).
- Replace T5 bulbs in germination chamber (every 1-2 years).

Notes:

POWER OUTAGE PROTOCOL (WITH CISTERN)

In the event of a power outage

1. Ensure backup oxygen system is running. Check flow meters to tanks and CFB.
2. Due to drain down effect, confirm water overflows to cistern.
3. Turn off main pump to prevent possible run-dry when power comes back on.
4. Determine cause of outage. If due to a circuit overload, consult electrician to correct.
5. Periodically check oxygen flow to tanks and CFB.
    - Watch for fish gasping at the surface and use D.O. meter to check levels
    - Adjust flow meters as necessary. Use small increments.

When power comes back on

1. Turn on main pump, restart flow to tanks.
2. Monitor sump level until system water is fully recharged. Replace water lost to drain down effect from cistern or source water.

Notes:

WATER FLOW ALARM PROTOCOL

If a water flow alarm occurs

1. Confirm backup oxygen has automatically activated. Check flow meters to tanks and CFB. Turn on manually if required.

2. Check sump level. Low sump is the most common reason for the problem.
- If sump is low, clean trough drain filters and fill sump from cistern or source water.
- Other common reasons for low sump are excessive SPA flushing or CFB filters are clogged.

3. If low sump is not the problem:
- Confirm main pump is operating. Check tank inlet spouts.
- If main pump is not running, proceed to Step 4.
- If main pump is running, proceed to Step 5.

4. Check for power failure and circuit breaker.
- If a popped circuit breaker, reset and assess cause of problem.
- If a power failure, refer to Power Outage Protocol.

5. Check for a clogged pump.
- Turn off pump, close valve and remove the suction pipe coupling.
- Visually inspect the inside of the pump and remove any debris.
- Prime suction pipe prior to re-attaching.
- Open valve before turning on the pump.

6. Once pump is operating normally, alarm system will send "return to normal" message.

7. Confirm backup oxygen is off. Turn off manually if required.

8. Monitor sump level until system water is fully recharged. Replace water lost to drain down effect from cistern or source water.

Notes:

Marketing and Sales

THIS BOOK IS DESIGNED as a guide to understanding, building and operating a commercial aquaponic system. The marketing of fish and plants is mostly identical to traditional aquaculture or vegetable farms with some aquaponic advantages. In this chapter, we will briefly touch on the subject of sales.

Aquaponic Advantages

As an aquaponic farmer, you have several advantages in marketing.

Year-round Production

For most customers (end consumer, retail or wholesale) in temperate or colder regions, purchasing locally grown foods all year-round, particularly greens, is a fantasy. Aquaponics makes this a reality.

There is a huge marketing advantage in producing quality local greens all year. In our locale, for at least four months of the year, we are quite literally the only local farm with any sort of leafy greens. This not only means our lettuce is in high demand and that we can charge a premium price, it also helps to establish a loyal customer base for the season when traditional farms offer similar products.

Increasingly restaurants and wholesalers are prioritizing locally sourced products and promoting these to their customers. Offering such businesses a year-round supply is a substantial advantage over traditional, seasonal farms.

Ethical Plants

Aquaponics provides year-round production using a tiny fraction of the water consumed in traditional farms and none of the chemical fertilizers required by most hydroponic farms. You can use this to your advantage by educating customers. People who care about buying local will usually care about the ethics of production.

Ethical Fish

Wild fish populations are massively overfished, with many species at risk of extinction. Using our system, you will raise high-quality trout in a system with no impact on wild populations. Raised correctly, your trout will be tender, pink and the most ethical choice for locally minded customers, which will create high demand.

Aquaponics Is Sexy

It's modern, it's technologically advanced, it's sexy. It's a human-created ecosystem that produces the highest quality plants and fish with low environmental impact. This is a big marketing advantage.

Display the unique qualities of your farm proudly. Include it in your logo. Display photos of your farm at markets. Highlight it on your website. And don't underestimate how big a draw it is at a farmers market to have both greens and fish displayed.

Potential Markets

There are four primary markets to consider: farmers markets, restaurants and retail stores, wholesale distributors and farm gate sales. Each have advantages and disadvantages. The common balancing factors are the price you can charge versus ease and volume of sale.

Farmers Markets

Farmers markets are the bread and butter of small-scale farmers. There is likely no market where you will receive a higher price per unit. Farmers markets vary widely depending on the location. Some major cities have numerous markets, often of varying quality, and it can be difficult to obtain a spot. Smaller towns may have only one market. Most markets run seasonally (typically May to October in BC) but some are year-round. Year-round markets will be a huge advantage for you in non-peak seasons.

The downside to a farmers market are:

- Requirement for specific equipment such as tent, table and signage.
- The considerable amount of time involved (length of market plus setup, tear down and travel).
- Customer service requires being "on" for long periods of time.
- Sales are not guaranteed and will vary from week to week depending on the season, weather and competition.
- Market fees (drop-in or season).

Depending on the vitality of your local market(s), downsides notwith-standing, we recommend farmers markets as your first choice for sales. There you will access customers directly and receive 100% of the revenue from your products. Any links added to the sales chain will greatly reduce your price per unit. The same customers may pay the same amount for your produce from a secondary vendor, but the percentage to you will be much less.

Tips for selling at farmers markets:

- If you are not an extroverted person who can easily engage with strangers, hire someone to run your booth. Having the right salesperson can dramat-ically increase the number of sales.
- Smile and engage everyone. Many customers are shy and need to be invited in for a closer look at your booth.
- Strive to have the same salesperson at the market every week as customers will build a bond with them.

Our farmers market booth with salesperson extraordinaire, Bre.

- Be consistent with attendance. Attending the market regularly is key to establishing regular customers. Strive to miss as few markets as possible.
- Be consistent with products. Variety is great but strive to have core products (head lettuce, salad mix and watercress for us) available each market day.
- Presentation matters. Have a large banner made to hang on your tent. Make your booth attractive and inviting. Present your products well (remember our method for keeping lettuce crisp and unwilted in hotter/drier seasons).
- Remember the adage: "stack 'em high and watch 'em fly." When building your display, create artificial height by placing crates or boxes underneath the display.
- Abundance sells. As your products sell, keep your tables looking full by condensing and compacting the display.
- Highlight aquaponics. In most markets, you will be the only aquaponic producer. Make a big deal of this to draw people to your booth. Highlight it on your banner. Display pictures of your system. Invite people for farm tours. Use the unique nature of your farm to your advantage.
- Think value added. Living lettuce sells for more. Small bags of herbs (e.g., watercress) sell for a premium. Bring your fermented fish fertilizer to sell.

Restaurants and Retail Stores

Next to farmers markets, restaurants and retail stores are likely the best option for sales: the price per unit is usually good, and items are typically sold by the case.

We define a retail store as any business that sells directly to end customers, adding one link between you and the customer. This contrasts with wholesale distributors who sell to retail stores and restaurants, adding two links between you and the customer. In our experience, restaurants and retail stores are very similar markets. Both are likely to be individual business locations, and both tend to pay similar prices.

When considering restaurants and retail stores, the key is delivery. You can spend a lot of extra time delivering to individual locations. We suggest creating a delivery schedule of one or two days a week to minimize driving time. Consider distance before approaching a business and apply a delivery charge if necessary. There are numerous businesses an hour or so from our farm who would like to buy our products, but it makes no sense to deliver that far away when we can sell everything locally.

Typically restaurants and retail stores will have set prices for different products which may vary seasonally and are often negotiable. Use your year-round production capacity to your advantage in negotiations.

Tips for selling to restaurants:

- Consistent supply is mandatory. Most restaurants have a stable year-round menu, and it is a big problem if their supply of a staple item like lettuce misses a week, forcing them to find an alternate source. You have the advantage of offering year-round supply but planning will be required to meet ongoing contracts.
- Offer consistent products, as you will likely supply a set menu. The occasional new product may be useful for a special, but the bulk of your sales will be the same products again and again.
- It's all about relationship. Sell yourself and your farm as much as your products. Good chefs want to build relationships with good farmers.

Tip for selling to retail stores:

- Variety is best. In contrast to restaurants who have more fixed needs based on set menus, an ever-changing selection is beneficial for many retail stores. This is not, in our opinion, justification for you to grow a wide array of products unless retail stores are your primary market and variety is required.

Wholesale Distributors

Wholesalers serve as intermediaries between farmers and retail stores and restaurants. They do not produce products and do not sell them to the end customers. They are an extra link in the supply chain.

With very few exceptions, you will make less per unit, often much less, by selling to wholesalers. The counterpoint to this is that most wholesalers will take as much supply as you can produce, often by the pallet, and will pick up from you. It is the least work and by far the least profit.

If you live in a reasonably sized town or any city, you should be able to find more profitable markets. We do not recommend selling to wholesalers unless you have exhausted the potential to sell to more profitable markets. Wholesalers can be used as a backup to move large quantities of unsold products, but your goal should be to avoid them.

Farm Gate Sales

Selling directly from your farm seems like an excellent idea: you can be open whenever you want, and customers come to you. But being "open" means being available and ready to sell at any time or operating on the honor system, and customers may be unwilling to make a special trip just to buy greens.

In our experience, the only times it makes economic and time sense to sell from the farm are when people are already there. For us, this typically means during our monthly tours. Almost everyone who attends our tours buys some of our products. We do not offer farm gate sales at other times.

Market Comparison

The easier it is to sell your products and the larger quantity you can sell to one customer, the lower the price you should expect per unit. It is up to you to determine what combination of markets fit your output, lifestyle and skills.

When setting prices, research the local market, including what other local producers are charging at markets and the prices at retail stores. Keep in mind that many customers will pay a premium for fresh locally produced food.

The following table details the prices we are paid or have been offered during market research.

	Farmers Market	Retail/Restaurants	Wholesale	Farm Gate
Lettuce head	$3.00–4.00	$1.50–2.00	$0.50–0.75	$3.00–4.00
Living butterhead	$4.00–5.00	$1.50–2.00	unknown	$4.00–5.00
Romain hearts 1 lb bag	$6.00	$2.00–3.50	unknown	$6.00
Salad mix ½ lb bag	$5.00–6.00	$4.00	unknown	$5.00–6.00
Cress 6 oz bag	$5.00	$2.50	unknown	$5.00
Trout (per lb)	$8.00–9.50	$7.00	unknown	$8.00–9.50

The RCA Sales Model

In our local city of Duncan on Vancouver Island, there is a well-established, well-attended year-round outdoor farmers market. We sell there every week that we have sufficient supply (roughly 45 weeks per year), and it constitutes our primary produce sales. We sell most of the remainder of our produce to local restaurants.

We sell all of our fish whole. We sell the majority, if not all, of each cohort to one local restaurant. Our trout is a staple item on their dinner menu (see photo on cover). When we have excess, we sell them directly to end customers who pay a little more per unit but not enough to justify the extra work for individual sales. We always have a wait list for our fish.

We have also sold to an organic delivery service (which falls under "retail store" despite not having a physical store front). The price per unit is good, but they are located an hour's drive away (each way), so we have mostly phased them out as we can sell everything locally.

We explored the option of wholesale distributors but ruled them out due to the low price per unit. Despite the ease of working with wholesalers (they pick up and will take everything), for us the low unit price is not tenable.

Promoting Your Farm

Depending on your local market, your capacity as a salesperson and your objectives in farming, you may or may not need to actively promote your farm. In our case, we have more demand than we can supply, and thus we do very little to promote ourselves as farmers in order to drive sales. We do promote ourselves as educators.

If you choose to promote your farm, you are likely to garner substantial interest from local media and groups whose focus is sustainability. Reach out to local newspapers and TV — they will be interested.

Keep an active social media presence. Posting updates on outlets such as Facebook, Instagram and Twitter will keep your customers engaged and spread news of your operation.

You will likely receive frequent requests for tours of your facility. We suggest holding tours at regular set times rather than individual bookings. Tours can be paid or free. We offer free public tours once per month and also regularly host school and community groups.

We recommend a gate and a large "we are closed" sign at the entrance to your farm, particularly if you can see the greenhouse from the road. Without a gate and sign, uninvited curious guests will often wander into the greenhouse unannounced, which can occupy a lot of time.

Creating a Business Plan

W E STRONGLY SUGGEST CREATING A BUSINESS PLAN as the first step in becoming an aquaponic farmer. It is not possible for us to provide an accurate business plan due to the large variables in building, operating and marketing. This chapter will lay out the parameters necessary to create a business plan for your farm.

Construction Costs

The cost to construct the system has the greatest variability, the biggest of which are acquiring property, labor costs, site preparation, choice of greenhouse structure/covering and access to power.

Property Acquisition

This is the single biggest variable. If you already own an appropriate site, this figure may be zero. If you are planning on purchasing a site, this figure is as variable as the price of land. Note only the upfront purchase price is considered here. Mortgage payments or rent are not included in this figure. They are in Operational Costs.

Labor Costs

Other than property acquisition, this is likely the biggest variable. The difference between doing most of the construction yourself and hiring contractors will be measured in tens and possibly hundreds of thousands of dollars. Building an aquaponic facility is labor intensive, and while many of the jobs are not overly complex, they are time-consuming and require a high level of skill and accuracy. As a rule of thumb, if you are hiring contractors for all construction jobs, expect the cost of the facility to at least double.

Site Preparation

Depending on the proposed location, site preparation can be as quick as 1–3 days or can be a major project costing tens of thousands of dollars. Choosing a location with a suitable site that requires minimal excavation is strongly recommended.

Greenhouse

A polycarbonate greenhouse costs as much as three times the cost of a double-layer poly greenhouse (est. $75,000 vs $25,000). A used greenhouse can likely be acquired for less than $10,000 but may have substantial downsides and risks.

Power

Many sites will require a new or upgraded power service to run an aquaponic farm. Depending on the location, this could cost as little as a few thousand dollars to tens of thousands if in a more remote area.

List of Initial Costs

The following is a list of costs that should be determined, as closely as possible, in your business plan. Some entries will require separation into multiple sub-entries (e.g., you may require multiple excavations). Small items that are likely to cost less than $100 (e.g., the heater in the germination chamber and a 50–60 gal. barrel for the Tank Manifold) are not listed individually here but expect them to cost several thousand dollars in total.

The following list is roughly organized from potentially highest to lowest cost:

- Property acquisition (down payment on property purchase)
- Contractor costs:
  - General Contractor (GC)
  - Carpenter
  - Plumber
  - Electrician
  - HVAC (for heat pump and walk-in cooler)
  - Greenhouse installation team
  - General labor
  - Aquaponic consultant, design and/or installation
- Greenhouse, 120′×36′ (or 40′) including heater, circulation fans, roll-up sides, etc.
- Site prep:
  - Excavation of site, Waste Tank area, sump, perimeter drain, cistern

- ▪ Various fills (gravel, drain rock, sand) including delivery
- ▪ Hauling of excavated material
- ▪ Support rings for fish tanks and Tank Manifold
- ▪ Water access to greenhouse site
- Power installation (100 amp service to greenhouse site)
- Fiberglass fish tanks, 3 @ 8′ diameter, 1 @ 4′ diameter
- Air-to-water heat pump
- Electrical components: panel, cables, outlets, boxes, solenoids, relays, etc.
- Plumbing and aeration components: pipe, fittings, valves, etc.
- Wood products (mostly 2×3s, 2×4s and plywood)
- Walk-in cooler
- Greenhouse installation:
  - ▪ Pier augering
  - ▪ Concrete for piers
- HID lamps/ballasts: 12 @ 1000W, 7 @ 400W
- UV sterilizers
- Tools required for construction
- LDPE, 20 mil @ 14′ wide
- Radial Flow Separator, 36″
- Polystyrene, 8′×2′ (min. 130 sheets; recommended min. 160 sheets)
- Light movers and rails (12)
- Building supplies (screws, bolts, staples, etc.)
- Air stones
- Aeration pumps (2)
- Water pumps (2)
- Shade cloth
- Cistern
- Stainless steel counter/sink
- Dissolved Oxygen meter
- Building permit fees
- Aquaculture license fees
- IBC totes (Waste Tanks)
- Monitoring system
- Hot water system
- pH controller
- Backup oxygen system
- Epoxy resin
- Paint
- Net pots, min. 7,500
- Seedling trays

- Totes
- Scales (2)
- Tanks covers
- Washing machine
- Chest freezer
- CFB media: screens and MBBR
- Fishing nets (4)
- Cleaning brushes
- Knives and/or scissors
- Salad spinner
- Small items (less than $100)

Assuming no property acquisition cost, it may be possible, as a very rough estimate, to build our design with new components (notably fish tanks) for as little as US$100,000 if you are a professional contractor with multiple skill-sets (e.g., carpentry, excavation and plumbing) and the capacity and time to do most of the construction work yourself — in other words, with virtually no labor costs and with no unforeseen snags.

Realistically, in most cases where you can assist with much of the construction work and provide free or cheap general labor but will hire contractors to lead each construction job, we would expect our design to cost between US$150,000 and US$250,000, not including land acquisition.

In extreme cases, whether due to all contractor work and labor being hired out at full price or unusual site problems, the design could cost more than US$250,000.

Ongoing Operational Costs

Operational costs are also highly variable, notably labor. Our system is designed to be farmed by 1–3 people at less than 40 combined work hours per week (not including sales, marketing and administration). If all labor is done by the principle farmers as it is intended (no hired labor), this will greatly change the ongoing operational costs.

Unless you have a mortgage or pay rent for your site, likely the next most variable expense is power. For example, the price per kW hour can be more than four times as expensive in Boston as in Montreal.

The following list is roughly organized from potentially highest to lowest monthly costs:

- Labor, hourly/weekly/monthly
- Mortgage payments or rent
- Ongoing modifications, maintenance and repairs (higher in first few years)

- Water management supplies:
  - pH controls
  - Testing kits
  - Mineral supplements
- Power
- Insurance
- Propane refill and tank rental
- Planting media (e.g., coco coir, coco chips, rockwool)
- Cohorts of fish (250 fingerlings every four months)
- Fish feed
- Seeds
- Farmers market fees
- Tools
- UV bulb replacement (once per year)
- Pesticides and fungicides
- Plastic bags
- Licensing and membership fees
- HID light replacement (large expense every 5–10 years)
- Double-layer poly replacement (large expense every 4–5 years)
- Oxygen tanks refill
- Promotion (business cards, web hosting)
- Office supplies (printer ink, paper, receipt books, pens)
- T5 bulb replacement (every 1–2 years)
- Bank fees and interest

Income Estimates

Much like construction and operation, it is similarly challenging to estimate income. The key variables are: (1) the type of plants you grow; (2) your success at growing those plants in terms of quality; (3) the number of harvested plants, determined by the weeks to maturity and percentage of plants grown successfully; and (4) the price you receive per unit.

For ease of estimation, we can eliminate the first variable by assuming you will only grow head lettuce, sold as whole units. This is not practical in reality, as we recommend variety for better sales and a percentage of your plants will be inglorious and thus used in salad mixes.

The second variable (quality) is addressed here by only factoring in units that are of high enough quality to be sold as whole units.

The third variable (quantity) is addressed by showing a range of production times (from 4–6 weeks) and a subset percentage of successful production. Yields above 90% are unrealistic.

The fourth variable (price) is shown as an average per unit. The lowest price ($0.50) assumes all heads are sold cheaply to a wholesaler. The highest price ($3.00) assumes selling a very high percentage direct to end customers for top dollar, such as at farmers markets, and selling the remainder to retail or restaurants for a good price. If you live in a city with a strong local market, produce exceptional quality and are a skilled salesperson, your average price per unit may reach $2.50 or higher.

In most cases, we suggest conservatively estimating your average price at between $1.50 to $2.00, factoring in multiple types of market and seasonal variability, and an average annual production time of 5 weeks with 20% losses or unsold products.

Value-added products, such as living butterheads or smoked fish, sell for a premium and can greatly increase profits. This chapter does not factor in such value-added products, nor does it factor in secondary activities such as selling fertilizer, raising pigs on crop residue or having an auxiliary garden or orchard fertilized by fish effluent, all of which can increase revenue.

Remember as well that your first year of production is gradual and takes twelve months or so to reach full capacity, assuming no major problems. The table below is based on full production capacity (starting year two). This is important for a cash flow projection.

Lastly, you should assume that at best you will break even on raising and selling fish. Typically, fish turn a profit only on a much larger scale than you will be producing. For this reason, fish income is not included in this chapter. Optionally, you can remove ongoing fish expenses (cohorts and fish feed) from the business plan and estimate that expenses and income will be approximately equal. Do not remove fish income and expenses from a cash flow projection, as income is considerably delayed from expenses.

Income Estimate Table

The following tables are a rough guide only to show the variability of income based on production times, production success and skill at marketing. The tables show estimates for 4-, 5- and 6-week production times, and at various percentages of crop sold (from 60–90%).

As an example, the highlighted cell in the second table shows a gross income estimate based on a 5-week production time (in the troughs) with 80% of the harvestable sites producing mature, quality plants (20% losses or unsold) that are then successfully sold for an average of $2.00 per unit. The formula for this cell is:

6,192 (harvestable sites) / 5 (weeks) × 0.80 (80%) = 990 plants sold

990 (plants) × $2.00 (per unit) = $1,980 gross weekly income

4 Week Production Time

	Average $0.50/Unit	Average $1.00/Unit	Average $1.50/Unit	Average $2.00/Unit	Average $2.50/Unit	Average $3.00/Unit
@ 60%	464	929	1,392	1,856	2,320	2,784
@ 70%	542	1,084	1,626	2,168	2,710	3,252
@ 80%	619	1,238	1,857	2,476	3,095	3,714
@ 90%	696	1,393	2,088	2,784	3,480	4,176

5 Week Production Time

	Average $0.50/Unit	Average $1.00/Unit	Average $1.50/Unit	Average $2.00/Unit	Average $2.50/Unit	Average $3.00/Unit
@ 60%	372	743	1,116	1,488	1,860	2,232
@ 70%	433	867	1,299	1,732	2,165	2,598
@ 80%	495	990	1,485	1,980	2,475	2,970
@ 90%	557	1,114	1,671	2,228	2,785	3,342

6 Week Production Time

	Average $0.50/Unit	Average $1.00/Unit	Average $1.50/Unit	Average $2.00/Unit	Average $2.50/Unit	Average $3.00/Unit
@ 60%	310	619	930	1,240	1,550	1,860
@ 70%	361	722	1,083	1,444	1,805	2,166
@ 80%	413	826	1,239	1,652	2,065	2,478
@ 90%	462	923	1,385	1,846	2,308	2,769

Final Thoughts

MODERN AQUAPONICS is still in its infancy. Like all pioneering ventures, getting in on the ground floor has unique potential for great rewards but also comes with great risks. Becoming a commercial aquaponic farmer will mean that you, like us, are not just following but actively expanding the breadth of knowledge and helping to move farming forward. We urge you to take time to weigh the risks and rewards carefully before breaking ground.

For those who opt to join us, we welcome you with open arms and look forward to collaborating with you. We are available to serve as consultants, and we offer custom designs of all sizes for both DWC and tower systems. We can be found at our website: www.raincoastaquaponics.com.

This is an exciting time to be a pioneer in farming. Good growing and good health to you!

Adrian Southern & Whelm King

Resources

Raincoast Aquaponics — our website **raincoastquaponics.com**

New Society Publishers — our publisher, source of numerous excellent books **www.newsociety.com**

Bright Agrotech — makers of the only drip towers we recommend **brightagrotech.com**

Pentair Aquatic Eco Systems — primary source for aquaculture equipment **pentairaes.com**

West Coast Seeds — supplier of all of our seeds **westcoastseeds.com**

Jefferson Solenoid Valves **jeffersonvalves.com**

Michael Timmons 3-day Aquaculture course **fish.bee.cornell.edu**

Cayuga Aqua Ventures — best source to purchase *Recirculating Aquaculture Systems* by Timmons & Ebeling. **c-a-v.net**

Sensaphone Web600 — monitoring system **sensaphone.com/products/ sensaphone-web600-monitoring-system.php**

Aqualarm.net — flow switch **aqualarm.net/cooling-water-flow-c-2**

Bluelab pH controller **getbluelab.com**

Dwyer Instruments — pressure switches **dwyer-inst.com/Product/ Pressure/SinglePressure/Switches**

American Netting and Fabric — shade cloth **americannettings.com**

OMRI **www.omri.org**

Pesticide Properties Database **sitem.herts.ac.uk/aeru/ppdb/en/**

Penn State Greenhouse IPM **extension.psu.edu/pests/ipm/agriculture/ greenhouse/greenhouse-manual**

Sources

Bradley, Kirsten. "Aquaponics: A Brief History." **milkwood.net/2014/01/ 20/aquaponics-a-brief-history**

Brechner, Dr. Melissa, Dr. A.J. Both and CEA Staff. *Cornell Controlled Environment Agriculture: Hydroponic Lettuce Handbook.* Cornell University, Ithaca, NY. **cornellcea.com/attachments/Cornell%20CEA% 20Lettuce%20Handbook%20.pdf**

Davidson, John and Steven T. Summerfelt. "Solids Removal from a Coldwater Recirculating System: Comparison of a Swirl Separator and a Radial-flow Ssettler." *Aquacultural Engineering,* Vol. 33, no. 1 (June 2005). **sciencedirect.com/science/article/pii/S0144860904001049**

Davidson, John et al. "Evaluation of Depuration Procedures to Mitigate the Off-flavor Compounds Geosmin and 2-methylisoborneol from Atlantic Salmon Salmo Salar Raised to Market-size in Recirculating Aquaculture Systems." *Aquacultural Engineering,* Vol. 61 (July 2014). **sciencedirect.com/science/article/pii/S0144860914000545**

"Feasibility Assessment of Freshwater Arctic Char and Rainbow Trout Grow-out in New Brunswick." Commissioned by the New Brunswick Department of Agriculture and Aquaculture, April 2010. **www2.gnb.ca/content/dam/gnb/Departments/aaf-aap/pdf/ Publications/Aqu/ArticCharrRainbowTrout.pdf**

Greenhouse IPM Manual with an Emphasis on Biocontrols. PennState Extension. Pennsylvania Integrated Pest Management. **extension.psu. edu/pests/ipm/agriculture/greenhouse/greenhouse-manual**

Holmgren, David. "12 Principles of Permaculture." **justlists.wordpress.com/2010/01/14/principles-of-permaculture**

Klontz, George W. *Manual for Rainbow Trout Production on the Family-owned Farm.* University of Idaho Department of Fish and Wildlife Resources. 1991. **aqua.ucdavis.edu/DatabaseRoot/pdf/TROUTMAN.pdf**

Morgan, Dr. Lynette. "Nutrient Temperature, Oxygen and Pythium in Hydroponics" *Simply Hydro,* article 4-3. **simplyhydro.com/nutrient_temp.htm**

Mousavi, Seyyed Alireza et al. "Effect of Carbon Source on Acclimatization of Nitrifying Bacteria to Achieve High-rate Partial Nitrification of Wastewater with High Ammonium Concentration." *Applied Water Science,* Vol. 7, no. 1 (Aug 2014). DOI: 10.1007/s13201-014-0229-z

Nick, Savidov Ph.D: "Evaluation and Development of Aquaponics Production and Product Market Capabilities in Alberta." Ids Initiatives Fund Project #679056201. August 17, 2004. **backyardaquaponics.com/Travis/Evaluation-and-Development-of-Aquaponics-Production-and-Product-Market-Capabilities-in-Alberta.pdf**

Schmidt, Larry et al. "Environmental Assessment for the Use of Hydrogen Peroxide in Aquaculture for Treating External Fungal and Bacterial Diseases of Cultured Fish and Fish Eggs." Upper Midwest Environmental Sciences Center, for USGS, June 2006. **fda.gov/downloads/Animal Veterinary/DevelopmentApprovalProcess/EnvironmentalAssessments/UCM072399.pdf**

Shammas, Nazih Kh. "Interactions of Temperature, pH, and Biomass on the Nitrification Process." *Journal (Water Pollution Control Federation),* Vol. 58, No. 1 (Jan. 1986). **jstor.org/stable/25042841**

The Aquaponics Doctors. **theaquaponics doctors.com**

Timmons, Michael and James Ebeling. *Recirculating Aquaculture Systems, 2nd Edition,* NRAC: Cayuga Aqua Ventures, 2002.

"UV Information." **pentairaes.com/uv-information**

"Why Matrix Media?" **brightagrotech.com/products/matrix-media**

Woynarovich, András et al. "Small-scale Rainbow Trout Farming." FAO Fisheries and Aquaculture Technical Paper 561. Rome:2011. **fao.org/3/a-i2125e.pdf**

Yue, Stephanie Ph.D. "An HSUS Report: The Welfare of Farmed Fish at Slaughter." Humane Society of the United States. **humanesociety.org/ assets/pdfs/farm/hsus-the-welfare-of-farmed-fish-at-slaughter.pdf**

Zheng, Dr. Youbin, Siobhan Dunets and Diane Cayanan. "Greenhouse and Nursery Water Treatment Information System: UV LIGHT." University of Guelph, Ontario, Canada.

"ZipGrow Tower FAQ." **brightagrotech.com/hydroponic-farming/ zipgrow-tower**

Glossary

Air lock: a bubble of air trapped in a pipe that restricts or prevents water flow.

Ammonia oxidizing bacteria: bacteria that oxidize ammonia into nitrites.

Array: a set grouping of tower rows in a tower-based system.

Autotrophic bacteria: derive their energy from inorganic compounds (e.g., ammonia and CO_2).

Bacterial Surface Area (BSA): the amount of surface area available for colonization by bacteria.

Bolt: rapid and typically undesired plant growth when vegetables set seeds.

Cohort: a group or batch of fish of similar age.

Combination Filter Box (CFB): a unit designed specifically for the RCA system that contains ten filter screens and an MBBR.

Cross brace: a horizontal truss that connects and strengthens the two halves of a greenhouse arch.

Deep Water Culture (DWC): a hydroponic production method in which plant roots are suspended in nutrient-laden water.

Denitrification: absorption of nitrogen by plants.

Dibbler: a tool used to make a hole in soil.

Double-layer poly covering: two layers of flexible polyethylene, sealed around the edges and inflated by a blower.

Drain Down Effect: the release of retained water in the system in a power outage or pump failure.

Effluent: water containing waste.

Endwall: a vertical wall at the end of a greenhouse.

Extruded polystyrene ("styrofoam"): synthetic polymer made from the monomer styrene consisting of closed cells.

Feed Conversion Ratio: the capacity of a livestock animal to convert feed into body mass, measured as a ratio of feed given to weight gained.

Fingerling: a small, young fish with scales and working fins.

Fry: the smallest, youngest fish capable of feeding themselves.

Head: the amount of work a pump must do to overcome gravity, vertical height and friction in the pipe.

Heterotrophic bacteria: derive their energy from organic compounds (e.g., fish feces and excess feed).

HID lamp: High Intensity Discharge lamp, typically High Pressure Sodium or Metal Halide.

Hydroponic Surface Area (HSA): the total surface area devoted to plant growth.

Hygrometer: instrument for measuring humidity.

IBC tote (Intermediate Bulk Container): a plastic tank or tote with an integrated pallet, designed to be moved by a forklift.

Inglorious plants: smaller, damaged or deformed plants that cannot be sold as whole units but can be used in salad mixes.

Kerf: the width of a saw blade.

Low Density Polyethylene (LDPE): thermoplastic made from the monomer ethylene. Can be flexible or rigid at different thicknesses.

Manifold: a pipe or chamber with multiple branches.

Meniscus: the curve in the upper surface of a liquid caused by surface tension.

Moving Bed Bio-Reactor (MBBR): a liquid space filled with biological filter material that provides a large BSA (Bacterial Surface Area).

MSDS (Material Safety Data Sheet): a document cataloguing information about a chemical or product.

MWP (Main Waste Pipe): the pipe into which the four fish tanks and RFS drain to remove effluent from the facility.

Nitrification: oxidation of ammonia by bacteria.

Nitrite oxidizing bacteria: bacteria that oxidize nitrites into nitrates.

OMRI (Organic Material Review Institute): organic certifier of products, notably pesticides and fungicides.

PAR (Photosynthetically Active Radiation): the spectrum of light available to plants.

Polycarbonate covering: corrugated panels of varying layers and thicknesses.

Pump curve: a graph showing the flow of a pump at varying head.

Purlin: a horizontal structural member that connects the arches of a greenhouse to each other.

Pyrethrins: botanical pesticide derived from the chrysanthemum plant.

Radial Flow Separator (RFS): a passive conical settling vessel with no moving parts that removes large solids from liquid.

Ridge vent: an adjustable passive vent running the length of the greenhouse roof.

Roll-up sides: adjustable passive vents at the bottom of the greenhouse side walls that are raised/lowered with a crank.

Shade cloth: woven polyethylene netting that blocks light transmittance by a specified percentage.

Source water: water supply used to fill and top up system water.

Supernatant: the mostly clear liquid at the top of a settling chamber.

System water: recirculating water that is pumped from sump to tanks and which then flows back through filtration and troughs to the sump.

Tank Manifold: the mixing vessel into which the fish tanks drain.

Tare: reset a scale to zero to compensate for the weight of the weighing container.

The Golden Ratio: a ratio of the total grams of per feed per day and the total HSA ($G^t/M^2/Day$) used to balance the system's inputs and waste removal capacity.

Towers/tower system: a hydroponic production method consisting of vertical tubes, typically square or round, filled with a media into which plant roots grow.

Acronyms

BSA	Bacterial Surface Area
CFB	Combination Filter Box
DWC	Deep Water Culture
HSA	Hydroponic Surface Area
MBBR	Moving Bed Bio-Reactor
MWP	Main Waste Pipe
RCA	Raincoast Aquaponics
RFS	Radial Flow Separator
SPA	Standpipe Assembly
UVT	Ultraviolet Transmittance

Index

About the Authors

ADRIAN SOUTHERN is an aquaponic farmer and entrepreneur from the Cowichan Valley on Vancouver Island, BC. After founding and operating an intensive urban farm and managing a local farmers market, he is currently steeped in all things aquaponics. After years of system perfection, he founded Raincoast Aquaponics and raises trout and vegetables for a living.

WHELM KING is a business manager, project manager and entrepreneur. He has worked and consulted in a wide variety of fields including the arts, agriculture, publishing, media and law. He pursues his passions of food sustainability, ethics and rock climbing in Nanaimo, BC.

A Note about the Publisher

New Society Publishers is an activist, solutions-oriented publisher focused on publishing books for a world of change. Our books offer tips, tools, and insights from leading experts in sustainable building, homesteading, climate change, environment, conscientious commerce, renewable energy, and more — positive solutions for troubled times.

We're proud to hold to the highest environmental and social standards of any publisher in North America. This is why some of our books might cost a little more. We think it's worth it!

- We print all our books in North America, never overseas
- All our books are printed on **100% post-consumer recycled paper,** processed chlorine free, with low-VOC vegetable-based inks (since 2002)
- Our corporate structure is an innovative employee shareholder agreement, so we're one-third employee-owned (since 2015)
- We're carbon-neutral (since 2006)
- We're certified as a B Corporation (since 2016)

At New Society Publishers, we care deeply about *what* we publish – but also about how we do business.

New Society Publishers
ENVIRONMENTAL BENEFITS STATEMENT

For every 5,000 books printed, New Society saves the following resources:[1]

46	Trees
4,205	Pounds of Solid Waste
4,627	Gallons of Water
6,035	Kilowatt Hours of Electricity
7,644	Pounds of Greenhouse Gases
33	Pounds of HAPs, VOCs, and AOX Combined
12	Cubic Yards of Landfill Space

[1]Environmental benefits are calculated based on research done by the Environmental Defense Fund and other members of the Paper Task Force who study the environmental impacts of the paper industry.